DRAGON ORB

Aurora

MARK ROBSON

SIMON AND SCHUSTER

SIMON AND SCHUSTER

First published in Great Britain by Simon and Schuster UK Ltd, 2009
A CBS COMPANY

1 3 5 7 9 10 8 6 4 2

Simon & Schuster UK Ltd
1st Floor
222 Gray's Inn Road
London
WC1X 8HB

A CIP catalogue record for this book is
available from the British Library

ISBN 978-1-84738-448-5

Typeset by Rowland Phototypesetting Ltd,
Bury St Edmunds, Suffolk
Printed and bound in Great Britain by
CPI Cox & Wyman Ltd, Reading, Berkshire

For April.
This is the one you've been waiting for –
your very own story.

Acknowledgements

Two real pilots from The Great War are mentioned in this book – Rittmeister Baron Manfred Von Richthofen (The Red Baron) of Germany, who was arguably the greatest flying ace of his time, and the Canadian flying ace, Squadron Leader Roy Brown RAF, who was initially credited with shooting him down. No disrespect is intended to the memory of either of these two individuals, (or their abilities) in this work of fiction. But given the setting for the Dragon Orb series and the nature of the story, the final flight of The Red Baron was too momentous an occasion to ignore.

Also, many thanks to Ms. Herminder Birdi for lending her expertise on eye conditions, and to Ms. Sam Cushing for her information about blast trauma and Epley's manoeuvre.

Contents

Chapter One
The Orb of Vision

Life after death from death before life,
Enter the new age, through deadly strife.
Greatest of orbs is – dragon's device.
Gifted for ever: life's sacrifice.

'Why do you let the night dragon suffer?' the voice whispered in Aurora's mind. 'Are you so cruel that you will not heal her?'

'Heal her? I have no ability to heal. I am a dawn dragon, not a day dragon,' Aurora responded indignantly. 'Who is it that accuses me of letting a fellow dragon suffer? Show yourself.'

Aurora slowly scanned the great chamber with her huge, amber eyes. Nothing moved. She and the night dragon, Shadow, were still alone in the hall of mirrors. Since becoming separated from the rest of

1

the party, Aurora and Shadow had waited while the others had moved deeper into the Castle of Shadows searching for the third of four orbs they required to complete their quest. Aurora had lost contact with Firestorm and Longfang, the two dragons sealed on the other side of the deadly mirror along with the four dragonriders, including her own rider, Elian. Where were they now that she could not reach them? Were they separated by distance, or by something more sinister? The mirror that separated the two dragons from their friends looked fragile, but Aurora knew better than to try to break it because it was coated with dragonsbane, rendering it impossible for the dragons to break through without risking a horrible death.

The hallway was still and silent. Her companion, the night dragon, Shadow, was curled in a circle, dozing.

It was a strange coincidence that the dragons should be split into their current pairings. This was not the first time that she and Shadow had become separated from Fire and Fang. They had last been forced apart a week ago after being chased by Segun and the senior night dragonriders and now, having only been reunited a few hours ago, already they were separated again. It felt almost as if fate were trying to keep the pairings away from each other.

Their journey from the Oracle's cave to this desolate sea fortress in the north of Orupee had been harrowing. Dragonhunters had hounded them the entire way. Aurora and Shadow bore multiple wounds from the hunters' spears and were weary through lack of rest. Shadow had suffered most. A nasty injury to one of her main wing muscles had made torture of every wingbeat. The day dragon, Firestorm, had healed Aurora's injuries earlier, but he could not heal Shadow. A day dragon's healing fire did not work on night dragons.

'I was a dragon once, a long time ago,' the voice continued in Aurora's mind. 'A dusk dragon – the leader of my enclave for many years. Is it true that dawn dragons can no longer heal? I find that hard to believe. You were ever the most gifted ones, blessed with the abilities of day and night dragons, yet also having special powers unique to your kind. We dusk dragons have never burned with the fire of the sun. Nor can we harness the silence of the night. Ours was ever the shadow gift. But you – you glow with inner fire. Why do you not use it to heal your companion? I sense she is in much pain.'

'Even if I could breathe healing fire, why should my efforts be any more effective than those of a day dragon?' Aurora asked. 'Besides which, I have no reason to trust you. This place appears nothing more than a gigantic trap, designed to kill dragons and their riders. Show yourself and I shall be better placed to decide if I should trust you.'

There was a slight pause. Aurora sensed the owner of the whispering voice considering her request.

'I cannot come to you now,' the voice replied. 'It is too late. Your friends have nearly reached their goal. I must await them here in my chamber if I am to fulfil my destiny. I have waited a very long time for my final release. You will have to trust me. You cannot breathe healing fire – of course you can't! But the fire is within you.'

Aurora snorted with indignation. If she had fire burning inside her, she would know it. The idea that she could somehow not feel the heat of fire inside her body was nonsense. But the voice did not stop. Its whispering words continued.

'You glow with the fire constantly,' it said. 'Your powers draw from both day and night, and are aligned to both. Unfortunately, we dusk dragons were not similarly blessed. Day and night dragons are polar opposites. Their abilities have never been compatible. Your powers, however, can be used on both. The only dragon your healing cannot touch is a dusk dragon, for we are your polar opposites.'

Aurora thought about this. What the voice was saying made a bizarre sort of sense, but still she was no wiser as to how she could help Shadow.

4

'What must I do to use this healing power you say I have?' she asked.

'I can't say exactly,' the voice admitted. 'I only witnessed a dawn dragon healing once, and it looked as though she concentrated her inner fire on the point of physical contact she had made with the other dragon. I must go. Your friends are here. My time has come.'

The voice faded. Aurora did not know what to think. Did she dare to try healing Shadow? What if it was another trap and her efforts added further hurt to the night dragon? She reached out with her mind once more, straining to reach her rider, Elian. It was no use. He was beyond her range. Fang and Firestorm also failed to respond. If they were in the same place as the owner of the voice, how had he spoken to her when she could not reach her friends? So many questions and so few answers – what should she do?

Shadow twitched and shifted her body position. Aurora watched the night dragon struggle to minimise her discomfort as she sought to rest. She was in awe of Shadow's resilience. Despite the obvious agony of her wounds, the night dragon had persevered without complaint. Aurora had never met another dragon with such strength of mind and body.

Moving closer to the huge black dragon, Aurora closed her eyes and began to concentrate.

'What are you doing?' Shadow sounded irritated. 'I want to sleep.'

'Relax,' Aurora told her. '*I want to try something that might help you rest more effectively. It should only take a moment.*'

Rather than risk messing with Shadow's most painful wound, Aurora decided to try her experiment on one of her lesser injuries. The oldest was a deep cut to the night dragon's shoulder that Firestorm had cauterised with his hottest fire to prevent Shadow from bleeding to death. The resulting burn had left an ugly mess of distorted and bubbled scales around the original injury. Aurora doubted it would ever heal properly on its own. Leaning close, she moved her neck around until it made gentle contact against the wound. Drawing her inner light, she tried to focus on the point of contact.

Shadow flinched away from her touch as if burned.

'What *are you doing, Aurora?*' Shadow asked again, her voice suddenly suspicious and more alert.

'*I'm sorry,*' Aurora replied. '*Did I hurt you? I didn't mean to.*'

'*No, you did not hurt me, but the sensation was strange.*' Shadow paused a moment. '*My shoulder feels cooler. It has been burning with the heat of Firestorm's* breath since we left the Valley of the Griffins. What did you do?'

'I'm not quite sure. I may have the key to healing your wounds, but I need to experiment a little more. May I try again?'

'Where are the dragonhunters?' Pell demanded. 'What are they doing?'

Pell's voice was like the irritating buzz of a fly in Kira's ear as her vision soared out from the chamber, high up into the air above the Castle of Shadows. Despite many dangers and traps, she and her three fellow questors, Elian, Nolita and Pell, had won through to the innermost chamber of the castle and gained possession of the third of the four dragon orbs – the Orb of Vision. The ghostly dusk dragon that had guarded the special plinth on which the third orb had formed had vanished now, leaving the four riders and two of their four dragons with the dilemma of how to escape from the castle and carry the orb to the Oracle.

Kira found that, with concentration, she could control the Orb of Vision and, by channelling its power, she could now see outside the castle. Using it made her feel as if she were staring through a window, except the window opened into the world wherever she willed it, and could be moved with a single thought. It was bizarre, yet the sense of power it gave sent thrills of pleasure through her.

She pressed her fingers harder against the orb, splaying them around its smooth, glowing surface to gain maximum contact. Energy surged into her and her mouth curved instinctively into a smile.

'Let's see what this thing can do,' she breathed.

'Be careful, Kira.' Elian's voice was distant, but his urgent warning somehow penetrated the trance-like state that the orb created. 'Remember what the ghost dragon told us. The orb might betray us.'

Kira's mind froze, poised to hurl her vision across Areth to her home village in the Racafian savannah. Elian was right. She needed to exercise caution. Try something easier first. She had already seen Segun, the power-hungry night dragon leader, flying away from the castle. Spying on him would be a useful test of her control. Relaxing slightly she sent her newfound power of vision whizzing across the sky until she located him still flying south and west on his dragon, Widewing. Zooming in closer, she circled them.

Segun's face was set in hard lines. His dark brows were knitted into a frown and his eyes stared with icy fury into the distance. She could almost feel the intensity of his thoughts. It was clear he did not intend to return to the Castle of Shadows, yet was resolved to prevent the four questors from completing their task to restore the Oracle.

A sudden change in Segun's expression made Kira withdraw. She flashed away, soaring high and fast back towards the castle. Somehow he had become aware of her. A shudder ran down her spine. It was eerie. One moment Segun had been staring straight ahead, the next his focus had shifted and he had looked right at her.

What had he seen? She thought the orb allowed her to travel in a totally disembodied state, but now it appeared this was not the case. Segun had definitely seen something, though Kira had no idea what. Elian had been right to urge caution. Had the night dragonrider seen enough to deduce the four questors had gained the third orb?

Swearing, she mentally berated herself for gifting him with this information. She paused high above the treetops and looked down. The great forest spread below her, a vast textured carpet across the land, and she watched as the tiny distant figure of Widewing flew in the direction of central Orupee and the Oracle's cave. Any thoughts Kira had enjoyed of spying on the night dragons at will were gone. It appeared likely that Segun knew what to look out for now and he would warn his men.

Deciding to be more careful, Kira zipped back

across the sky towards the Castle of Shadows. Diving down until her vision was skimming the treetops, she slowed as she approached the edge of the forest. She could see the dragonhunters gathered outside the great castle gates. She lowered her viewpoint down amongst the branches and then eased forwards again until she was peeping out from between the leaves at the open ground in front of the enormous sea fortress. The hunters were some distance away. They had split into two groups – one to either side of the main gates. It appeared they were preparing an ambush.

Wrenching her fingertips from the surface of the orb, Kira staggered backwards. As soon as her contact with the orb was broken, she began to see through her own eyes again. The chamber seemed to spin under her feet. After the amazing perfect vision she had just experienced, her own sense of sight felt very limited. Also, the disorientation she had felt when Longfang, her dragon, sacrificed his left eye to form the orb, returned. The sensation was hard to define, but the dragon's loss of perception echoed through their bond making her feel dizzy.

'Are you all right?' asked Elian. 'What did you see?'

It took a moment for Kira to regain her composure. 'Segun's not coming back,' she confirmed.

10

'I followed him a little way. He's heading southwest towards the Oracle's cave. The hunters, however, have laid a trap for us directly outside the castle gates.'

'That's not good,' Pell observed. 'We can't take off from inside the walls. There's not enough room for a take-off run and those shadow demons in the main courtyard will rip us to shreds if we make any sudden movements.'

'Do you think the hunters know about the shadow demons?' Elian asked thoughtfully. 'If their lead hunter is a Joining, possessed by a demonic creature, would he be aware of the shadow demons inside the castle?'

The four riders each looked around at one another. No one had an answer.

'What do you think, Fang?' Kira asked, keeping the tone of her thought curt. She was still angry with him for sacrificing his eye to form the third orb. Fang had solved the Oracle's riddle before they had reached the inner chamber, but had kept the information from her to prevent her getting upset.

'I don't know,' he replied. *'Demons, as you call them, are not from this world. I do not know the extent of their powers. They may be aware of one another, but I don't even know if the creatures that join with humans come from the same world as the shadow demons. The Joining*

11

does seem to have a sense unlike that of any human or dragon. He has tracked us and anticipated our moves with uncanny accuracy. Anything is possible.'

'Fang doesn't know,' Kira said aloud.

'Neither does Firestorm,' Nolita added softly.

'I wonder . . .' Elian began thoughtfully. Pacing back and forth next to the metal plinth, he hooked his right thumb under his chin and curled his index finger across his face between his top lip and his nose.

'What, Elian?' Kira asked. 'What do you wonder?'

'I was just thinking that perhaps we could set a trap of our own,' he said. 'But first we need to get out of this chamber and find our way back to Aurora and Shadow. Something is stalking the maze of mirrors. I'd rather not go back that way if we can avoid it. Let's concentrate on getting out of the castle. I'll tell you my idea once we're in a position to act.'

Chapter Two
Surprise Attack

Tembo felt tension building inside his body as he watched Husam, leader of the dragonhunters, directing the other hunters into position on either side of the castle gateway. It was a sensation that had become all too common over the past two weeks. What was Husam thinking? It would not matter where the men were placed. It was too open here in front of the fortress. Had he not learned from their foolhardy attempt to finish off the two dragons on the ridge five days ago?

The open attack up the ridge had ended in disaster when the two dragonriders had used their position of strength on the high ground to great advantage. One of the dragons had rolled entire tree trunks down at the hunters, whilst the riders had hurled a deadly barrage of rocks. Tembo eyed the battlements on the walls high above them. If the dragonriders became aware of the hunters again, the walls of the fortress could be used to give a similar advantage.

Assuming the dragons could not take off from within the walls, however, the gateway would funnel the dragons when they left, allowing the hunters to launch a targeted attack. But as the ground in front of the castle was so devoid of cover, the hunters would be as exposed and as vulnerable as the dragons.

A bitter wind swirled around Tembo, increasing his tension. He flexed his legs. His muscles were stiffening. They had chased the dragons for more than a hundred leagues over the past week, and his body was not used to being still for any length of time. The men would not want to be out here for long. There was something else, though. He could not shake the nagging feeling that they were missing something.

Barely an hour ago, a particularly large night dragon and his rider had launched from the treeline less than five hundred paces away from them. Husam had been as unaware of the dragon's presence as the rest of the hunters. Given their leader's uncanny instincts and tracking skills, Tembo had been surprised to see Husam caught off guard. It was not the same night dragon they had been tracking. This

14

one was huge and had flown with no hint of any injury.

The beast was airborne and out of reach before anyone had a chance to react. If the rider had gone for help, Husam and the hunters were as good as dead already. If the entire night dragon enclave turned to tracking them down, then there would be no hiding place on Areth secure enough to guarantee their safety.

Only the strongest and best of their party had dragonbone weapons now. Most of the bone-tipped spears had been lost during previous encounters. Those without dragonbone weapons were reduced to carrying blades and spears tipped with metal, which would be next to useless if things turned nasty. Husam had outlined a bold plan, but it was fraught with danger. He was counting on the dragons being weakened by their wounds and the effects of the long chase. If Tembo had learned anything from these past two weeks, it was that the dragons they had been hunting were full of surprises. Their riders were young, but they were no fools. Husam's plan lacked subtlety. The dragons were sure to see through it.

Tembo shifted uneasily as he crouched, waiting. Apparently satisfied that his men were suitably positioned, Husam turned and strode towards him.

Slim and upright, the hunter moved with a bounce in his step.

'Are you alright?' Husam asked.

'I'm just weary,' Tembo sighed, hoping his friend would not probe further. He could not look Husam in the eyes. No one looked him in the eyes by choice. There was something about Husam's mismatched irises that chilled the marrow from one's bones. Both eyes had been blue, but after the disastrous attack that had resulted in the death of their former leader, Kasau, one had darkened to a colour that was almost purple. The other remained a more normal shade. It was strange, but not as strange as his friend's change in behaviour.

'Don't lie to me, Tembo,' Husam said, his tone dangerous. 'You're thinking my strategy is flawed. We're too exposed. We have limited weaponry. If dragons come from anywhere but through the castle gateway we'll be totally at their mercy. You're right to think these things.'

The admission was a surprise. Tembo looked up at Husam and instantly regretted it. The man's eyes glittered with anger. His gaze seemed to jab into Tembo's head.

'Trust me, my friend,' Husam ordered. 'We will leave this place triumphant today. I *know* we will.'

Tembo had never been argumentative. Despite his

16

misgivings, he nodded and climbed to his feet. He towered over Husam, but his size meant little in this relationship. Husam had always been the leader.

'I have always trusted you, Husam,' Tembo said slowly, though in his heart he knew that to be a lie, too. 'But this is as far as I go. If we don't kill the dragons today, I'm giving up the hunt. This has to end. There'll be other dragons: dragons without riders; legitimate rogue dragons that need destroying. I'd like to hunt them with you, but I'll go alone if need be.'

Pressure began to build inside Tembo's head. Husam's stare intensified. Pain flared and Tembo clamped his hands over his temples. He could not close his eyes. He could not even blink. Husam's eyes filled his mind. He had no idea what Husam was doing, but he was determined to hold firm to his decision.

The next few moments felt like an eternity, but suddenly Husam broke off his stare. Tembo staggered slightly before regaining his balance.

'Very well, Tembo,' Husam said. 'But you need not fear. It will end today. We have been through a lot together, you and I. It would be a shame if we had to part company now.'

Tembo rubbed at his temples with his fingertips. Kasau had played a similar mind trick when the hunting party had first latched on to the trail of the dawn dragon. Husam's attempt to persuade him by force of mind had been less subtle than that of Kasau, but it had felt very similar. Husam was not the man Tembo had taken up with in Racafi. He had known it in his heart from the moment his friend's eye had changed colour, but he had so wanted to believe the change in Husam was temporary that he had convinced himself to persevere as if nothing was wrong.

'Husam!'

The hissed whisper from one of the nearby hunters was urgent.

'Husam. The watchman!'

Tembo followed the man's gesticulations and realised the watchman poised by the side of the gate was signalling to them. The dragons were on the move. This was it.

Husam waved a rapid sequence of silent instructions to the men. Tembo watched as the men reacted to the signals. Their response was impressive. The watchman withdrew from the gateway. Everyone was poised within moments, weapons at the ready. He could feel his heart pounding with anticipation and he began counting the beats to help keep a sense of calm. *One hundred . . . one hundred and fifty.* Where were the dragons? Had the watchman been mistaken?

Two hundred. A movement to Tembo's right drew his attention. It was Husam. He was creeping towards the watchman's position at the side of the great arched entrance to the fortress. Tembo's breath caught in his throat. What did Husam think he was doing? If the dragons were coming, he would be virtually under their feet as they emerged. Without a miracle, he would suffer the same fate as Kasau. To attempt to fight a dragon at close quarters was to invite death.

Husam passed the watchman and moved silently along the wall until he was right next to the gateway. Twisting, he took a quick peep into the castle and then instantly flattened himself back against the wall. He remained still for a moment and then he turned and took another look. This time he moved more slowly, easing into position and staring in through the gateway.

The signal to gather was a further surprise. What was Husam up to now? What had he seen? Tembo ran forwards as lightly as he could. The others in position nearby ran with him. Seven men gathered next to the wall on the other side of the gateway.

'Get ready,' Husam warned. 'We're going in.'

'I thought you said . . .'

'I know what I said,' Husam snapped softly through gritted teeth. 'But this is too good an

19

opportunity to miss. The dawn dragon is asleep in the courtyard. There's no sign of the riders. The night dragon is somewhere in the keep.'

Tembo shook his head. 'It feels like a trap to me,' he said slowly. 'Why else would the dawn dragon expose herself like this?'

Husam's eyes narrowed, but his expression was more thoughtful than angry.

'You might be right, Tembo,' he replied, his focus going distant. 'But I'm certain that if we act swiftly there will be no other dragons close enough to come to her aid before we make the kill. The injured night dragon will not bother us. There are some strange echoes inside the castle. I sense the dusk dragon and day dragon have rejoined the two we have been following. They are somewhere within the walls, but they are not close enough to cause us a problem. The dawn dragon is alone in the courtyard.'

Tembo knew better than to ask how Husam knew all this. He suspected the truth was better left unknown. The other men looked confident. Despite the long journey and the repeated failures to bring this hunt to a close, they appeared eager to follow their leader into the fortress without any apparent concern for what fate awaited them. Tembo did not know how Husam had won this unswerving loyalty from them, but it was another of those stones he felt were best left unturned.

'Come on, men,' Husam urged softly, his gaze never leaving Tembo. 'Let's go before the other dragons complicate matters.'

Signalling the men on the other side of the castle entrance to follow, Husam slipped around the corner and began creeping through the gateway. Tembo rolled his shoulders a couple of times to loosen them, hefted his spear in his right hand and joined the end of the line. If the others were mad enough to go in there, the least he could do was watch their backs for them.

No sooner had Tembo rounded the corner into the tunnel-like entrance of the sea fortress than he could see the bright golden orange of the dawn dragon ahead. She was asleep, curled tight with her eyes shut. Her scales gleamed with the promise of wealth. Tembo was not immune to their effect, but for every step he took looking forwards, he spent two checking behind him for any sign of the trap he sensed closing in.

The hunters passed under the portcullis and onwards until they reached the far end of the tunnel. Stopping short of the open courtyard inside the fortress walls, the men bunched together against the walls on either side of the gateway.

Tembo watched as Husam scanned the area ahead and around. Even from the back of the short line, Tembo had a fairly good view of the courtyard. The only movements he could see were several faint swirls of dust devils where the strong wind was whipping over the battlements and twirling loose debris in tight little columns of air. Husam was right. There was no sign of the other dragons, or their riders.

Husam signalled again and the hunters began to spread swiftly in a wide arc, running lightly on silent feet. Those at the furthest ends of the lines were just reaching their positions when disaster struck. Tembo saw movement out of the corner of his eye and he froze.

One of the dust devils raced across the courtyard so fast that it was little more than a blur of swirling debris. To Tembo's horror the twister moved with purpose – straight towards the man at the end of the line. There was no time to call a warning, nor did the man cry out as the dust devil struck. The man vaporised as Tembo watched, disintegrating into a fine mist of red. His weapon toppled to the ground with a clatter. There was nothing else left to show that he had ever existed. Terror locked Tembo's muscles. He could not move.

More dust devils began to move, gliding swiftly

across the stone paving of the courtyard. Panic struck the hunters. They turned and ran back towards the castle gateway, but the twisters were too fast. One after another, the men disintegrated as they were caught by the twirling nightmares.

Husam was nearby. He was also standing motionless. The two of them were nearest to the exit and the only ones not attempting to escape. Tembo tried to open his mouth, but his jaw would not respond. He wanted to cry out and urge Husam to run, but he found he could not do anything.

Tembo's eyes were drawn to another of the dust devils. It was coming straight towards him and approaching fast. Tembo closed his eyes and waited for it to strike. His body tensed tighter, bracing for the impact. But it did not come.

His breathing sounded loud in his ears and his body tingled all over. He could feel a presence in front of him. It began moving. Circling around him. Nothing physically touched him, but Tembo felt that even with his eyes tight shut he could 'see' where the thing was at all times. Another joined it. Then another. He could feel them all around him. Why did they not kill him? He did not dare open his eyes to find out.

They circled for some time before they began to move away. Still Tembo kept his eyelids firmly clamped shut. It was a gurgling sound that finally tempted him into opening them. Unable to move or speak, he was shocked to see Husam apparently suspended in mid-air. He was upright with his feet about waist height off the ground and he was shaking like a fish, freshly dragged from the water. Several of the dust devils surrounded him. They looked like swirling air, but were really living creatures. As Tembo concentrated, the hint of an outline teased the corners of his vision. The shape that formed was so horrible and improbable that his mind instantly rejected it.

What were these things doing to Husam? Why had they not killed him and Husam in the same way they had killed the other hunters? It made no sense.

A sound began to issue from Husam's lips – a high-pitched whine that resonated in Tembo's ears and set his jaw aching. It was hard to credit that a human voice could produce such a sound. Suddenly it dropped to a more normal register and became more recognisable as a scream of agony. Throwing his head back, Husam hurled his voice at the sky. Tembo watched. Horrified. Enthralled.

As Husam's chest emptied of air and his voice trembled to silence, a cloud of dirty grey smoke coiled from his open mouth. It hung in the air for a moment like a personal thunderhead, then a sudden gust of wind, stronger than any Tembo had felt through the day, reached down with invisible fingers and whipped the miniature cloud away. Husam's body instantly sagged. His arms went limp at his sides and his head lolled forwards until his chin came to rest on his chest.

So slowly that Tembo failed to notice it at first, Husam began to descend. His feet contacted the flagstones but the creatures did not drop him. They continued to lower him slowly – first his legs and then his body folding gently, until he rested, a crumpled heap on the ground.

Time had ceased to have any meaning. Tembo could not tell if he had been standing still for a minute or an hour. The swirling vortices moved away steadily until they reached their original locations around the castle courtyard. Still Tembo stayed motionless. Not until the creatures had remained still for some time did he begin to consider moving.

Heart hammering, he forced first one foot forwards, then the other. Step by step, he eased across the short distance to where Husam was curled. None of the creatures showed any sign of movement, so he reached down slowly, worked his huge hands under Husam's slender torso and lifted his friend into his arms.

Husam was not heavy, but Tembo's arms were

shaking by the time he entered the shadow beneath the huge arched castle entrance. Once clear of the courtyard he accelerated quickly, first lengthening his strides and then breaking into a run. He reached the open ground in front of the castle and turned immediately left. The mournful cries of the gulls wheeling in the air high above the battlements echoed his emotions as he placed his friend on the cold ground.

'Husam!' he said urgently, patting his friend's cheek. 'Husam, wake up! We need to get out of here.'

'Wh . . . what?'

'Husam! Are you all right? Talk to me.'

'I'm fine,' he muttered, his eyelids fluttering as they opened. 'What happened? Where the hell are we?'

'You're in northern Orupee,' a voice said from behind Tembo. 'And you've been hunting the wrong dragons.'

Chapter Three
Pardoned

Elian stood over the unconscious leader of the dragonhunters and his huge companion, with his sword held ready. The man's eyes looked around nervously as he weighed up the situation. His square face with its sausage of a nose and big ears that stuck out almost at right angles from the side of his head made him appear simple, but Elian could see the intelligence in his eyes. If Elian had been alone, the hunter would probably have tried to overpower him and make a run for it. But Elian was not alone. Kira and Nolita were right behind him. The girls were armed with knives, and the way they held them left the hunters in no doubt that the weapons were not for show. Pell and the dragons were now emerging from the castle as well. There was no chance of escape. The distance to the nearby forest was too great.

27

The hunter on the ground gave a groan. 'Northern Orupee?' he asked in a querulous voice. 'How did we get to northern Orupee?'

The big man looked down at his fellow hunter and gasped with surprise. 'Your eyes!' he exclaimed excitedly. 'Your eyes are the same colour, Husam!'

Husam looked even more confused. 'Of course they are, Tembo. My eyes have always been blue. You're confusing me with Kasau ... but Kasau is dead. At least I think he is. Everything is so hazy. What are we doing here?'

'You're yourself again, aren't you?' Tembo persisted, unable to contain his excitement. 'You led us here, but that wasn't the real you. You acted like Kasau – all mysterious and weird. And you knew things – when and where the dragons would appear. The rest of the men followed as if under a spell. I followed because I knew the man beneath the strangeness – the man I'm looking at now. It's great to have you back.'

Husam looked at Tembo as if he were mad.

'*Is he right?*' Elian asked Aurora. '*Is this man no longer a Joining?*'

'*I no longer sense anything unusual about his mind,*' Aurora confirmed. '*This is the man who led the hunters, but it appears the shadow demons forced the creature inside him to leave.*'

'Why would they do that?'

'Demons hate every living creature, Elian,' she explained. 'But most especially other demonic forms. Dragons have encountered many demons over the centuries, but what we have learned of them makes for uncomfortable knowledge. It seems likely that the shadow demons expelled the joining creature from this man's body to prove their superiority over it. They could not kill it in this world. It is my understanding that for one demon to kill another, both would have to cross the barrier into their world before the fatal blow could be struck. Crossing the boundary between worlds is torture for all demons. Weakening the Joining by reducing it to a bodiless spirit was the most damage they could deal it without an excruciating sacrifice.'

'But why not just kill this man as they did the others?' Elian asked. 'Surely if they had destroyed his body, the Joining would have been left bodiless in the same way.'

'Not quite,' she said, her voice thoughtful. 'By leaving the host body alive they added insult to injury. The Joining was sent to roam bodiless, whilst the host that had served it well was kept alive and unharmed. You can be sure the Joining would not be happy about that outcome.'

'And this other man?'

'Stayed still as we did. He showed the respect the demons require of those they let live.'

29

'*I see,*' Elian said, running his fingers through his hair as he contemplated the consequences of the shadow demons' actions. '*So what shall we do with the hunters, Ra?*' he asked.

'*Let them go.*'

'*Just like that?*' Elian was surprised. '*After all the hurt they caused you and Shadow?*'

'*Well if it were left to Shadow, she would probably eat them,*' Aurora admitted. '*But she's in a good mood now that her wounds are healed, so I don't think she will argue. The men are unlikely to trouble us again. Give them a warning and let them go.*'

'What are you talking about?' Husam was saying to Tembo. 'You're not making any sense.'

Elian listened as the two hunters began discussing what had happened.

'My dragon tells me you no longer pose us a threat,' he interrupted, making sure he had both men's attention. 'I'm not so sure.'

'Stick them with your sword and be done with it, Elian,' Pell said, his voice cold and emotionless. 'They tried to kill us enough times.'

Elian had not heard the older boy approach.

'Not so fast, Pell,' he snapped, unable to mask his irritation. 'Let me deal with this. I would rather not kill anyone unless it's absolutely necessary. There's been more than enough blood spilled today.'

'Let Shadow do it then. She's not squeamish.'

Elian glared at him. Pell returned the stare for a moment, shrugged, and then stomped off in disgust. The two girls remained where they were, knives drawn, a few paces behind Elian.

'Shadow, the night dragon over there, wants to eat you. She is rather angry about all the injuries you and your men dealt her,' Elian told the two hunters. 'My dragon, Aurora, is willing to try reasoning with her. The way I see it, I have several options – I could let Shadow have an early dinner; I could kill you myself; or I could hand you to the nearest Overlords and leave you to be executed for hunting dragons that were not rogues. But Aurora has offered a more lenient alternative. Will you listen?'

The two men nodded.

'Aurora has suggested that we let you go,' he continued. 'She feels enough men have died today. But if I release you, I will require you to hand over all of your dragonbone weapons, and I need to be certain that you'll never try to hunt us again. I'm sure you're aware of a dragon's ability to search a man's mind. I want you to promise me that if you hunt dragons in future, you will only seek to kill true rogues. Be careful how you reply. If my dragon senses even a hint that you are lying, we'll let Shadow have an early dinner.'

The two hunters looked first at each other and then at Elian.

'We promise,' they said together.

'What do you think, Ra?' he asked aloud.

'Their response was genuine,' she confirmed.

'My dragon is not sure,' he said aloud, keeping his face stern, and staring intently at first one hunter and then the other.

'I vow on everything I hold sacred that I will never hunt any dragon again unless I know it to be a rogue,' Husam said. He sounded on the edge of desperation.

'Me too,' Tembo said quickly.

Elian pretended to consider for a moment. 'Very well,' he said. 'You may go free.'

The men heaved a simultaneous sigh of relief.

'Can you stand, Husam?' Tembo asked. 'If you're all right, I'll check all the saddlebags for dragonbone weapons. I dropped my spear in the castle courtyard. The men took their weapons inside, but I'm not going back in there to retrieve them. I'd rather have the dragon eat me.'

'Sensible,' Elian agreed. 'Only an unpleasant death awaits you if you go back.'

'Go ahead, Tembo. Check the horses. I'll be fine,' Husam assured him.

'Are we finished here?' Kira asked. 'We need to get

going before Segun cooks up any more trouble for us.'

'Nearly,' Elian replied. 'Go and mount up. I won't be long.'

Kira and Nolita sheathed their knives and walked to where Fang and Fire were waiting. Pell was already sitting in his saddle, clearly eager to leave. Tembo ran across the open ground to the edge of the forest. It took a while, but Elian was surprised at how fast he moved for such a big man. He returned out of breath and shaking his head.

'There's nothing there,' he gasped. 'All the weapons that remain with the horses are made of metal. I swear it.'

'He is telling the truth,' Aurora confirmed.

'Very well,' Elian said, sheathing his sword. 'Remember my words, though. If word ever gets back to me, or any of my friends, that you have been hunting any dragon other than a true rogue, nowhere on Areth will be safe. We will hunt *you* down.'

Aurora had moved up close behind Elian. He turned and mounted her, bounding up her foreleg and swinging into the saddle with confident ease.

'Tell the others we're ready,' he told her. *'Let's see how far we can get before nightfall.'*

Aurora stared intently at the two hunters for a moment, slowly baring her teeth. Suddenly she snapped open her mouth and let loose an almighty roar that sent Tembo and Husam scrabbling backwards with their hands clamped over their ears. With that she whirled on the spot and sprang away across the open ground, spread her wings and launched into the air.

Elian concentrated on holding on as they left the ground. The wind swirling around the great castle was fickle, constantly changing its direction and strength. Aurora compensated instinctively, but the conditions did not make for a smooth take off. Even once airborne the air near the ground was turbulent, making his dragon dip and bump unexpectedly. It was only when they had climbed clear of that low-level choppy air that Elian began to relax more into the saddle and enjoy the flight.

The other three dragons flew in loose formation around them. Aurora turned south and east towards their distant goal of the Oracle's cave. They had done it again. Another orb successfully secured, but not without cost.

He remembered how upset Kira had been when she found out that Fang would have to sacrifice an eye to form the third orb. He wondered how he would feel if he and Aurora had had to make that decision. He shuddered, all too aware that it was his turn next. What sacrifice would the final orb require? The last line of the Oracle's verse had sent shivers down his spine from the moment he had first heard it. He hoped there was more to it than the obvious interpretation.

> Life after death from death before life,
> Enter the new age, through deadly strife.
> Greatest of orbs is – dragon's device.
> Gifted for ever: life's sacrifice.

'Thinking about our task?' asked Aurora.

'You know I am.'

'Do not fear,' she said, her tone warm and comforting. 'Our part of this quest will require no more than we can bear. The Oracle would not ask it of us otherwise.'

'You seem confident of that,' Elian noted. 'After seeing what Fang and Kira went through, I find it hard not to worry. Pell irritates me beyond reason sometimes but I agree with his questioning of the Oracle's motives.'

'I confess I was disturbed by our last encounter,' Aurora admitted. 'There was something strange about the Oracle that I did not like.'

Elian was surprised by the admission. For his dragon to imply anything but good of the Oracle was not something he expected to hear.

'I've got an awful feeling there is something that we're missing,' he said. 'The Oracle knows. I'm sure it does.

35

But it won't tell us. There is a mystery about the restoration of the Oracle that doesn't feel wholesome. Pell is right. We need answers. And I'm not just talking about finding the orbs.'

Aurora did not respond this time, but he could feel her presence in his mind. She was projecting a sense of peace through the bond, but his mind was not ready to accept her feelings. There were too many questions running amok for him to settle.

'Do you think Kira will ever forgive me for what I did to Fang?' he asked her. *'I feel awful for having helped remove his eye, but I knew it was what he wanted. If the final decision had been left to Kira, we wouldn't have the third orb. I can't say I blame her. If it had been you wanting to do it, then I'd probably have reacted in the same way.'*

'She knows you meant well,' Aurora said carefully. *'But your action hurt her and it is not in her nature to forgive quickly. Nor will she forget. It will take a long time for her to reconcile the events of today. I have already asked Fang the same question, but his answer was an evasion. You will have to be patient. She will come around.'*

Elian could see the wisdom in his dragon's words. Patience would be required. The problem was that they did not have time to be patient. They only had until the harvest full moon to find the final orb and deliver it to the Oracle. They had done well to get the first three orbs so quickly, partly because in recovering the first two orbs, they had gained clues about the location of the third. This time they had nothing. And in reciting the verse, it made no more sense to him than it had the first time.

Although Aurora was not flying very high, the wind was cold against Elian's right side. It was nothing like the freezing conditions in the mountains of Isaa, but this subtle cold was perhaps more dangerous. He had no intention of pushing anyone's limits. The Castle of Shadows had left its mark on all of them, and Kira, in particular, would be vulnerable after the traumas of the day.

The forest was slipping by beneath them, but the expanse of trees was so vast that for a while it seemed almost as if they were not moving at all. The sun had dipped considerably towards the horizon before they saw the first break in the trees. There was a village in the clearing, with enough clear land next to it for the dragons to land.

'*Fang is suggesting we stop the night here,*' Aurora told him.

'*I think that's a good idea,*' he said. '*We can barter for some decent food and possibly even get beds for the night. After weeks of sleeping on the ground, it would be great to have a proper mattress.*'

The dragons descended in a line. Firestorm led, followed by Shadow, Aurora and Fang. They landed on the open ground near the village. Elian felt his dragon pump her wings extra hard in the final few spans as the ground loomed large. The wind was curling over the treetops, creating localised rotors of air that made judging the landing a tricky business. Firestorm, Shadow and Aurora made good landings, but Fang landed awkwardly, thumping his taloned feet into the ground and staggering to an ungainly halt. Elian saw Kira lurch in the saddle, but she did not fall.

'*Are Fang and Kira all right*,' he asked Aurora anxiously.

'*I think Fang's pride is a little dented*,' she replied. '*But aside from that, they're fine. Landings are bound to cause him a few problems for a while. He has lost his depth perception and learning to adjust to his limited vision will take some time.*'

Dismounting, Elian intended to speak with Kira, but a small group of people from the village was already congregating nervously nearby. From old to young, all looked at the four enormous dragons with wide eyes. Fear, awe and excitement rippled back and forth across the group of faces and they spoke to one another in hushed voices.

Elian looked at his fellow riders and let out a soft

sigh of resignation. Kira was clearly in no mood to talk to the villagers, Nolita was already washing her hands with water from her bottle and Pell would probably manage to alienate these people within two sentences. Once again, it appeared he would be the ambassador for the quest party. He did not feel comfortable in the role, but he could see he had little choice.

Putting on his friendliest smile, he strode towards the village, opening his hands in the recognised sign for peace.

'Here we go again,' he breathed without moving his lips.

Chapter Four
Fishing

Kira's fingers tightened around the orb. As before, whilst using the orb, her sight felt as if it were detached from her body and had gone soaring through the sky until she was looking down from a tremendous height. This time, it had flown away to the mountains surrounding the Oracle's cave. Her sight climbed so high that almost the entire range was visible, laid out in a great sweeping line across the centre of Orupee like an army of rock giants.

Kira could not see the Oracle's cave from this angle, but she could see where it was situated. As she stared intently at the area around the cave mouth, she began to pick out the night dragons. From so high up, they were difficult to make out, but the longer she looked, the more dragons became visible. Her caution had paid off. She was not sure what Segun had seen when she had used the orb to spy on him before, but it seemed unlikely that the night dragonriders would spot anything at this distance.

Making a rough tally, Kira withdrew and returned to her body, placing the orb back into her saddlebag.

'Well?' Elian asked. 'Is Segun there? What did you see?'

'We're in trouble,' Kira replied, her lips tightening into a thin line. 'I counted at least fifty night dragons in the valley outside the Oracle's cave. I didn't get close enough to do an accurate tally, but once I'd got to about thirty I realised it was irrelevant. Segun has set up a blockade and it won't be easy to get through.'

'Fifty!' Elian gasped.

'I imagine there will be more coming, too,' Pell observed, his voice emotionless. 'He's probably sent word for the entire enclave to gather there. Unless we get in, he wins. The Oracle will die.'

'Can't you just get Fang to camouflage and sneak past them?' Nolita suggested.

'It's not that easy,' Kira said, shaking her head. 'That's exactly what Segun will be expecting. He knows our abilities now. That's why he's brought in so many riders. It looks as though he's given up on beating us to the orbs. Pell's right. He's attempting to set up an impenetrable perimeter around the Oracle's cave.'

'What about a diversion?' Pell suggested.

'What sort of diversion?' Kira asked. 'It needs to be spectacular if it's to have any hope of drawing away enough of the night dragons to make any difference.'

'I'm not sure yet,' Pell admitted. 'I thought maybe Elian and I could draw them off while you sneak in and deliver the orb.'

'You're not leaving me alone,' Nolita said, her voice cracking with a squeak of fear.

'No one's leaving anyone alone,' Elian said firmly. 'We need some time to think. If we make a wrong move now, the quest will fail. Whatever we do, it's likely that we'll only get one chance. The chances of all those dragons just happening to be in the vicinity of the Oracle's cave right now are too remote to be considered. Segun devised this blockade some time ago. He's thinking ahead. We need to do the same if we're going to outwit him. Let's set up camp while we make our plans.'

It was a good place for a campsite. The hulking peaks of central Orupee dominated the western skyline with their snowy caps, but the dragons would have to fly for an hour or more before they entered the range. They had landed in an idyllic green vale next to a small lake. The evening was warm and the air was alive with the sounds of insect life. There was a constant low-level buzz of flies, while butterflies fluttered on silent wings and an occasional bee meandered in dancing flight across the green surface.

A light breeze brushed through the leaves of a nearby stand of mature trees that offered plentiful resources and a great location to build their shelter. The water in the lake tasted fresh and harboured a potential source of food.

Elian noted rings on the surface of the water where fish were rising. He had a coil of fine cord in his pack that had been there since he left home, but he had not considered using it until now. His father had shown him how to fashion hooks the previous summer.

After the bleak windswept landscape around the Castle of Shadows and the freezing conditions of northern Isaa this place seemed like paradise. It had been easy to forget that although midsummer had passed, autumn was still some way off. The leaves were still a rich green, with no sign of turning. If it had not been for the pressing nature of the quest, Elian would have been more than happy to enjoy the weather and relax. The prospect of basking next to the lake with a fishing line in his hand was most appealing, but other priorities called first.

It took less than an hour for the four riders to build an impressive shelter under the trees. Armloads of ferns made for a soft bed under the cover of a lattice of branches. On top of the lattice they arranged more layers of ferns, with the tips pointing downwards and the shiny side of the leaves uppermost, to encourage any moisture to run down the pitched surface.

The sun was still high in the sky when Elian sat down at the bank with his coil of fishing cord and his assortment of swiftly fashioned wooden hooks. He selected one, tied it to the end of the line and baited it with a small piece of meat. He tied the other end of the line loosely around his wrist. It had been tempting to fashion a pole, but Elian had no intention of messing about playing with the fish. He would settle for hauling them onto the bank to cook over the fire later.

To his delight, the fish were all too happy to oblige. The first offering of meat was snatched before it had barely sunk below the surface. Elian gave a sharp tug to set the hook, but he was a little too enthusiastic – the hook was not strong enough to take the sudden load and he felt it snap.

'Damn!' he cursed.

'Problem?'

Elian started. Kira was standing right behind him.

He had not heard her approach. Even through the ferns, she moved without making a sound.

'I wish you wouldn't do that,' he grumbled. 'You're for ever making me jump.'

Kira flashed him a mischievous grin. 'It's my favourite pastime,' she said. 'Pell doesn't bite the way you do and Nolita is so twitchy that she usually sees me coming. How are you doing? I've never tried fishing.'

'Well if that first cast was anything to go by, I should catch something,' he replied, pursing his lips as he examined what was left of his hook. 'I just pulled a bit hard, that's all.'

'Show me.'

Kira sat down and watched as Elian cut the shaft of his hook from the line and tied another in its place. He baited it again, feeding the meat onto the hook so that the sharp point was free to do its work. Standing up, he whirled the bait on the cord and then released it in a lob that carried it a good way out over the surface of the water. It splashed down with a gentle *plop*, sending concentric ripples outwards from the point of impact.

Elian took up the slack in the cord, watching the line as the bait sank slowly. Again the bite came swiftly. This time Elian was more careful in his response. He gave a slight jerk on the line to set the hook into the fish's mouth and was surprised at the strength of the pull he felt as it responded. Whatever he had hooked was not small and it raced away beneath the surface, the line zipping through the water. Before he had a chance to think about what was happening he felt the second hook snap as well.

Elian was tempted to curse again, but he resisted the urge.

'Looks like you're going to need a stronger hook,' Kira observed innocently.

'Really?' he responded. 'You don't say!'

'Here,' she said. 'Try this.'

Kira pulled a little leather pouch from her pocket. Reaching carefully inside, she took out a metal hook. It had been fashioned with a tiny loop at the end of the shaft to make it easy to attach to a line.

'Where did you get this?' Elian asked, fascinated. 'It must have been incredibly difficult to make.'

'The man who made weapons for our tribe gave it to me as a gift before I left,' she said with a slight shrug. 'He liked to fish in the river near our village. I never thought I'd use it, but I'd be happy to watch you try.'

'That's amazing!' Elian said. 'I don't think I've ever seen such a delicate piece of metalwork. It's finer than a piece of jewellery. I'd be terrified I might lose it.'

'Don't worry,' Kira assured him. 'I'm not likely to miss it if you do. Go ahead. Let's see if you can catch something. I quite fancy fish for dinner tonight.'

Although for a piece of metal it was incredibly fine, the hook was considerably larger than the wooden ones Elian had fashioned. His father had told him that fish could be wary of a big hook but, given how fast they were biting, Elian did not imagine that would be a problem today. Taking care to fasten it securely to the line with a self-tightening knot, Elian baited the metal hook and stood up to make another cast. A minute later he lifted the first fish onto the bank. A sharp rap to the top of the its head with a large stone stopped it flapping around.

'It's twice as long as my foot!' he exclaimed excitedly. 'I don't think I've ever seen a fish this big before.'

Kira smiled at his enthusiasm. 'There's enough meat on it to feed us all,' she agreed. 'But if you could catch another, then dinner will be a feast.'

Elian was more than happy to try. It did not take long. The second fish was not quite as large as the first and he was tempted to throw it back and try again, but common sense prevailed. Even given the ease with which he had caught these fish, there was never any guarantee that the fish would continue biting.

Catching more than they could eat was wasteful so, with one last wistful look at the water, he lifted his knife to cut the metal hook from the cord.

'Keep it,' Kira said, gently placing a hand on his wrist to stop him from making the cut. 'You're more likely to use it than I am. I'm too much the hunter to be bothered with fishing. It's yours.'

'Really? But it was your leaving gift . . .'

'And now it is my gift to you,' she said firmly. 'I'm not blind, Elian. You enjoy the thrill of hunting the water as much as I enjoy stalking a deer. You'll gain far more pleasure from the gift than I will.'

'Thanks, Kira. I wish I could give you something in return. I'll do my best never to lose it.'

'I know you will, Elian,' she smiled. 'I think that's what I like about you. You're so practical.' She paused for a moment. 'I've been thinking about what you said earlier and you were right.'

'About what?'

'We'll only get one chance to get past Segun's blockade,' she said. 'One chance, not two. We still have to get the last orb, Elian. I'm not sure we should try to take the Orb of Vision to the Oracle just yet.'

Elian did not answer straight away. He finished coiling the line and wrapped it back into its leather wallet. Hooking a finger through the gills of each of the two fish, he picked them up and began walking slowly back up towards the campsite. Kira walked silently by his side. He could not help thinking that Kira might be addicted to wielding the orb. If so, then the longer she kept it, the more dangerous the orb would become.

'Go on,' he prompted finally. 'What do you have in mind?'

'I think it makes sense to find the last orb first and then run the blockade with both of them. The final orb might also have a power that we can use to help us.'

'It might,' Elian replied, his tone wary. 'Then again, it might not. I'm concerned about keeping the third orb any longer than necessary. You saw what the first orb did. I saw the effects of the second as well. Neither of the orbs brought good to their rider. I've got a horrible feeling that we've not really seen the full effects of what this orb can do yet either.'

'It helps us see at distance,' Kira said. 'What's wrong with that? I admit that I don't understand it totally, but the power it has is useful. Look at today, for example. Without the orb's power, we'd be flying into Segun's trap right now.'

'True,' Elian agreed. 'I don't deny the benefits. I'm just not convinced we've seen the full effects of carrying the orb yet. A chill runs down my spine every time I think of the ghost dragon that guarded the orb in the Castle of Shadows. He warned us against using the orb in case it betrayed us.'

Kira frowned. It was clear to Elian that she was not convinced about this. Did she not believe the ghost dragon had been telling the truth, or was the power of the orb seducing her?

'However,' he continued. 'I've also been thinking about Segun's blockade and I think Pell had it right. We need a diversion – a huge diversion. Pell and I could never hope to draw the attention of more than a handful of the night dragons. We're going to need help for this.'

'Help?' Kira asked. 'What sort of help?'

'Well, I was thinking that the combined help of the entire compliment of the day, dawn and dusk dragon enclaves would probably draw their attention.'

Kira stumbled and stopped. She looked at him with wide eyes.

'You're talking about all-out war!'

'Not necessarily,' he countered. 'I'm talking about a show of strength that would hold their attention while we sneak in and deliver the orbs.'

'Orbs?' Kira leaped on the word. 'So you agree with me?'

Elian nodded reluctantly. 'I think it's the only way.'

Chapter Five
Treading the Tightrope

'You've got to be joking!' Pell said, his eyes flashing with anger. 'I'm not going to babysit Nolita while you two go adventuring through the other world to find the last orb.'

The evening light was dimming fast, but the flickering orange flames from the fire illuminated the faces of the four companions. Despite the dragons settling down some distance away, Elian could see that the fear in Nolita's eyes was stronger than ever. He felt sorry for her, but he knew he could not back down in the face of Pell's attitude. He had waited until everyone had eaten their fill of fish before explaining his idea. His father, Raim, had always claimed people were more accepting of bad news when they had a full stomach.

'If we're going to succeed in this quest, then I don't

51

see any of us having much choice,' Elian replied, keeping his voice calm and steady. 'Nolita can't go to the day dragon enclave alone.'

'Perish the thought!' Pell muttered sarcastically. He stabbed a stick into the base of the fire and a flurry of tiny sparks leaped upwards, spiralling and dancing into the air like a cloud of fireflies.

Elian ignored Pell's comment. 'Of all the dragons, Fang is the best equipped to move through the other world safely,' he continued. 'The soldiers there won't shoot their stingers at something they can't see. We don't know where the last orb is. Our best chance of finding it is to visit the dawn dragon enclave and look for clues.'

'So you keep saying,' Pell said. 'But what makes you think you'll find anything there? I don't see anything in the final verse that leads us to want to look there:

> *Life after death from death before life,*
> *Enter the new age, through deadly strife.*
> *Greatest of orbs is – dragon's device.*
> *Gifted for ever: life's sacrifice.*

What's any of that got to do with the dawn dragon enclave? Look what happened to me when I went to my enclave for help.'

'No disrespect to your enclave, Pell,' Kira interjected, 'but I doubt we'll experience the same sort of welcome at the dawn dragon enclave that you received from Segun. The rhyme says the orb is a dragon's device. Maybe one of the dawn dragons has already made the final orb. We won't know until we ask. Also, we can request some of the dawn dragons to pay a visit to the dusk dragon enclave in Ratalucia. Fang ran rings around three night dragons in the snowstorm up in Isaa. Imagine the chaos a dozen or more dusk dragons could cause amongst Segun's ranks. The dawn dragons could bring the dusk dragons through the other world and have them here in Orupee in a couple of days.'

Pell's expression became thoughtful as he considered Kira's words. His brow remained furrowed, but his eyes lost some of their inner anger as he followed her logic through.

'I agree that having more dusk dragons to help us would be useful,' he conceded. 'But I still don't see why Nolita can't go to the day dragon enclave alone. It's her enclave and she's been there before. Or Kira could go with Nolita, and I could come with you. Nolita has flown Firestorm every day for the past few weeks. It's not as if I can hold her hand from Shadow's back. Besides, we know now that Aurora can heal Shadow if she gets injured, but she can't heal Fang. Putting day dragon with night and dawn dragon with dusk will weaken the partnerships. The abilities of the dragons clash, rather than complement.'

'Nolita can't go alone because none of us should fly alone at the moment,' Elian explained patiently. 'Both parts of this plan are equally important and cannot be entrusted to any one rider. If we don't find the final orb, the quest fails. Aside from Fang's abilities, Kira also has the orb of vision and I'd like access to the power it offers during my search. There might be a way of using it to find the final orb. Likewise, without the help of the day dragons we've got no chance of getting past Segun. If you want to give Segun a bloody nose, the best way would be to hit him with something hard. I'd say you could do worse than to turn up with a flight of several hundred fire-breathing day dragons at your side.'

Elian watched Pell across the fire. He could almost see the picture forming in the older boy's mind: a mass of blue dragons with a single black dragon at their head. Pell's ego would place him nowhere else. He was close to accepting the plan. Elian could see it in his eyes, but he also knew that Pell was like the fish he had caught earlier. He was not the sort to just roll over and allow himself to be hauled in. He would fight all the way to the end.

The biggest worry now was not Pell. It was Nolita. Leading her felt like walking a tightrope. One false step and it would all be over. She had come so far. Elian could not shake the feeling that the Oracle's message about working together held a significance that went beyond teaching Pell a humbling lesson for his rebellious trip to the night dragon enclave.

The blonde girl was hugging her knees, rocking back and forth gently and staring into the fire. She had said nothing throughout the exchange. When Elian had first mentioned going to fetch the day dragons he had noted that her initial expression held a mix of fear and excitement. Fear had dominated her features each time the possibility of her making the journey alone was raised. Nolita had successfully travelled with Kira all the way from the north of Isaa to the far side of Orupee. It was clear she had become comfortable in Kira's company and Elian felt a stab of guilt for separating them especially as he knew that she did not like Pell. Was he really sending Pell with Nolita because he thought the combinations made for the best teams? Or did he want to offload Pell because he preferred Kira's company?

For the briefest moment Elian thought about taking Nolita with him, but he dismissed the idea instantly. Sending Pell and Kira together was a recipe for disaster.

Kira was the most able leader of them all, yet she was also able to follow when the situation demanded. She was tough, both mentally and physically. With her painted face and her braided hair she looked, from head to toe, the hardened tribal huntress. The intelligence and strength behind her eyes lent confidence to those who followed her lead. She had some strange ideas at times, but Elian respected her skills and liked her straightforward approach. Although he knew her to be capable of subterfuge, any ulterior motives that Kira might have were subtler than Pell's.

'*Fire tells me that Nolita is close to despair,*' Aurora warned. '*She draws a lot of strength from Kira's company. Fire is worried. He says she is considering drastic measures to avoid having to go with Pell.*'

'What sort of measures?'

'*He would not say,*' the dragon replied. '*But I can tell from his tone that her plan would not help our cause. You had better talk to her.*'

Elian thought for a moment. He did not know what to say to Nolita, but he knew that Aurora was right. If he did not deal with her feelings now, she might do something silly – hurt herself, or worse.

'What do you think, Nolita?' he asked gently. 'You've not said anything yet.'

'Why would you care what I think?' she replied,

still staring into the fire. 'I'm just here to make up the numbers now.'

'Nonsense,' Kira said quickly. 'You're as much a member of this team as any of us. Of course it's important what you think.'

Nolita looked up from the fire and met the eyes of each of the others in turn before beginning to speak. 'I think we're in out of our depth,' she said eventually. 'I think that all paths are fraught with danger and that I should never have been chosen for this quest. I'm not suited to life as a dragonrider. I'm not comfortable with any of this. I don't want to fly to Racafi with Pell and I certainly don't want to go back into that other world again. Given my own choice I wouldn't fly anywhere – but you know that already.'

'I sympathise with your feelings, Nolita.' Elian reached across and placed a hand on her shoulder as he met her troubled gaze. 'But what do you think we should do?'

'I think we should give up on the quest,' she sighed. 'Segun is brutal and ruthless. He will kill us all rather than let us deliver the orbs to the Oracle now. He's got a mass of support from his fellow night dragonriders. Even if we do get the day dragon enclave to intervene, what good can they do against an army of night dragons?'

57

'She has a point,' Pell acknowledged. 'If it came to a straight fight, the night dragons would dominate. They are stronger and faster. Besides, I've been having second thoughts about restoring the Oracle for some time now. Let's just say for a moment we're successful and we find the final orb in time and somehow get past the blockade – what will happen when we throw the last orb into the Oracle's pit? I don't trust the Oracle. It's up to something. It's not been telling us the full story about these orbs. I get the feeling it's been playing us for fools ever since we set out on this quest.'

An uncomfortable weight seemed to settle in Elian's stomach. A strangely cold burning sensation crept up through his chest to the back of his throat as he recalled his last encounter with the spirit creature. Pell was right. The Oracle had not been totally honest with them. When Pell gave it the second orb, Elian had not felt at peace in its presence. There had been something in the Oracle's voice – a note of greed maybe? He had tried to put his concerns out of his mind for the sake of his dragon's life purpose, but doubts had been raised. What was more, he knew his dragon shared those doubts.

The crackle and pop of the fire seemed to get louder as an awkward silence settled. Other noises began to intrude on Elian's thoughts. Insects chirruped and whirred through the air on gossamer wings. The leaves whispered at the caress of the evening breeze. An owl screeched, the sound carrying across the water with startling clarity.

'The question we must ask . . .' Kira began. She paused until she was sure she had everyone's attention. 'The question we must ask is: What is the alternative? Let's say we accept that the Oracle has ulterior motives. I think the Oracle's agenda is unlikely to be as dark as Segun's. We need to think about what will happen if we abandon the quest now. We know something of Segun's intentions from Pell's experience in the night dragon enclave. As soon as the Oracle is dead he plans to launch an offensive. Is he serious, Pell? Does he really think the night dragons can dominate Areth?'

'Absolutely!' Pell said without pause. 'And he's probably right. Segun said the death of the Oracle will free his hand to seize ultimate power and rule the world. He called it "the age of dragons" and he's deadly serious.'

'Assuming we let the Oracle die and Segun embarks on his plan, what do you think he will do first?' she asked.

Pell looked thoughtful for a moment. 'He'll look to remove his most dangerous opponents from the field,' he said. 'The day dragons are the only enclave strong enough to give him any serious problems. He'll look to crush them swiftly. Once the day dragons are no longer a threat, he'll send night dragons throughout Areth with the same message – submit to his authority, or face the consequences.'

'That's what I thought,' Kira said, nodding. She turned and looked Nolita in the eye. 'We don't have a choice, Nolita. This isn't a game. Scary though it is, the fate of Areth is in our hands. If we fail, our families and friends could soon be answering to Segun. Is that what you want? Can you imagine the night dragons paying a visit to your family in Cemaria?'

Nolita shuddered and Elian found he could almost picture her thoughts as the images tumbled through her mind – the black dragons swooping down from the sky to the tiny village in the woods. The people there had no weapons that would be effective against a night dragon. She had told him that her sister and brother, Sable and Balard, were brave, but their bravery was likely to get them killed if they resisted Segun's men. He watched her face intently as she came to the realisation that Kira was right. Segun could not be allowed to see his plan through.

'I'll do it,' Nolita said softly. 'I'll go to the enclave with Pell. The thought of the journey scares me, but the thought of not acting scares me more.'

'Thanks, Nolita,' said Elian, his voice sounding both grateful and relieved. 'I knew we could count on you. You've been brave throughout this quest. Your family will be so proud when they hear about what you've done.'

'*I've* not agreed to go yet,' Pell pointed out.

'But you *will* go,' Elian said firmly. 'If we're going to get past Segun, we're going to need all the help we can get. Getting the day dragon enclave to support us is our best chance.'

'What if I decide I want to help the Oracle less than I want to hurt Segun?' Pell suggested. 'If I'm honest, I don't trust either of them. And I like them even less.'

Elian's eyes narrowed with anger. 'Sometimes I think you argue for the sake of it, Pell,' he muttered through gritted teeth.

Pell gave a short bark of laughter and shrugged his shoulders. 'That's a fair assessment,' he admitted. He poked at the fire again with his stick. 'But you're right. I will go. It will be worth the trip just to see Segun's face when the day dragons line up against him – assuming they will come, of course.'

'They'll come,' Nolita said, her eyes distant and her voice sounding strangely prophetic. 'They will come and Areth will never be the same again.'

Chapter Six
Dark Passage

'Look after her, Pell,' Elian ordered, keeping his voice to barely more than a whisper. 'I know how you feel about her, but she's braver than you give her credit for and I can't help feeling she's still got an important part to play in this quest.' He glanced across to the shelter. He could just make out the mop of blonde hair unmoving inside.

Pell's eyes instinctively followed Elian's. The older boy nodded. 'Don't worry, Elian,' he said. 'I'll make sure she gets to the day dragon enclave, but what she does then will be up to her.'

'Good enough,' Elian replied. 'When you get there, ask for Barnabas. He's the leader. He'll know how to handle Nolita. He'll also be able to organise the enclave. Ask him to have the day dragons in position by the day before the harvest full moon.

That gives you just over two weeks, which should be ample time to fly to the enclave in Racafi and back. Hopefully, Kira and I will be waiting for you with the last two orbs.'

'Where will you wait?'

'There's a lake about two hours' flight south of the Oracle's cave,' Elian said. 'Viewed from above, it looks like a grasping hand. It is the biggest for some distance around. You can't miss it.'

'We'll be there. Good luck with finding the orb.'

The pre-dawn glow was increasing steadily, as was the swell of the dawn chorus. Aurora was eager to get airborne. Elian could feel her energy and anticipation. Kira was already mounted and waiting. Leaping lightly up Aurora's side, Elian settled into his saddle. With a farewell salute to Pell, he gave his dragon the silent command to leave.

Aurora turned on the spot and bounded away through the green sea of ferns with Fang following close behind. Within moments they had left the ground and were climbing across the still water of the lake. The air was cool, but not cold, and the sky was cloudless as far as the eye could see. The breeze had died out, killing all hints of turbulence. It was a beautiful morning for flying.

Elian could feel the dawn approaching. He had become attuned to Aurora's instincts. He could feel her strength building and her will gathering as she prepared to open the gateway. The sudden flood of energy that poured through her raged and roared in a mind-numbing torrent, as she tore apart the fabric of time and space and opened the vortex. It swirled in front of them, waiting like the gaping mouth of a monster about to swallow its victim whole.

Kira and Fang surged past them and flew straight into the grim grey vortex at high speed. They vanished into the grey cloud, disappearing the instant they entered. Elian braced himself for the passage between worlds. The twisting wrench of entry felt as though his body had been distorted out of its natural shape. Even though he had made the jump to the other world several times, experience did not reduce the discomfort.

From the moment he entered, Elian could no longer feel Aurora under him. The familiar sensation of weightlessness was no longer as terrifying as it had been the first time. But the total lack of any sensation was as spooky as ever. No sound. Nothing to see. No smell. No taste. No sensations of touch or pressure on his body. It was good that it did not last long. Such a lack of stimuli could easily cause a person to lose their mind if they were subjected to it for any length of time.

The wrench of emergence was welcome, but any

thoughts of relief were short-lived as the air around him suddenly erupted with fire and noise.

'What on Areth . . .?' Elian gasped.

It was slightly darker here, but the explosions were so frequent and so close that the nearby countryside appeared more vivid than it would in full daylight. White and orange fire split the air around them and exploded from the ground below in great sunflowers of flame. Dawn was not far away. The eastern sky was a crown of light blue.

'We've arrived in the middle of a battle,' Aurora said urgently. *'Fang has started to climb. We need to get away quickly.'*

'No arguments here!' Elian replied.

The drone of several flying machines filled the momentary gaps in the barrage of explosions. There were a number of machines both ahead and behind them. What were they doing flying at this time of night? Unless they had launched some time ago, they must have got airborne in the dark! How did they keep from crashing into one another?

'The machines are fairly evenly spaced,' Aurora said. *'The men in the machine approaching from behind us are full of anticipation. They are preparing to drop weapons on the area directly below us. The ones in the machine in front are relieved. They have dropped their weapons and are concentrating on looking for landmarks to navigate*

home by. They seem to be following a set route in a steady stream.'

'What's down there?' Elian asked.

'Large buildings and some horseless carriages that are mounted on metal lines on the ground,' she replied. 'There are quite a lot people down there, too.'

'Is Jack in one of the flying machines?'

Elian felt Aurora reach out with her mind. She searched the sky around them looking for the airman they had encountered on previous visits to this world. All the time she was searching, she was also climbing higher and higher, her wings pounding in a powerful rhythm that lifted them out of the immediate path of the flying machines.

'I don't think so,' she said eventually. 'Though I can't be totally sure.'

Elian was also scanning the sky when a sudden blast of light, heat and sound smashed into him with no warning. By chance, he was looking to his right, directly at the spot where the weapon exploded. Instinct made him flinch away. As his head snapped to the left, he felt pain flash forwards in a line from his right earlobe across his cheek and a simultaneous flare of pain erupt in his right thigh. Aurora lurched into a turn to the left, roaring with anger and shared pain. The suddenness of the rolling motion combined with the shockwave from the explosion nearly threw Elian from the saddle. Only his instinctive reaction to grip the pommel with all his might saved him from a long fall into the dark.

The next few minutes were confusing. Elian's ears were ringing and his right ear was incredibly painful. Everything sounded distant and unreal. Nor could he see very much. A large yellow flash spot blocked his vision wherever he looked. Everything around the spot was blurred, and blinking hurt the surface of his eyes. Even with his eyes closed, the yellow splash of colour remained.

Aurora appeared to be having problems flying straight and her voice in his mind sounded distant. He could tell that she was also hurt. Her thoughts were laced with a rising sense of panic, but he could not think straight enough to respond with any coherent thoughts of his own. There were more explosions. Some were close enough to rock Aurora further, but none inflicted more pain.

An inner fire began to burn in Elian's mind and chest. Power was building inside him. The sun was approaching the horizon. In his confused state it took a while to realise the fire he could feel was Aurora gathering energy to form another vortex. He wanted to call out. He could feel that the full power of dawn was still some way off. He wanted to tell her to stop – tell her not to overstretch herself. The thought of entering a passage back to Areth while injured terrified him. Jumping between worlds was uncomfortable enough when he was fit and well, but at the moment he was barely hanging on to her back. What would happen if he fell inside the vortex?

He wished he could communicate with his dragon, but he could not channel his thoughts into any sort of organised stream. He wanted to ask how she was, but he could not reach her. He did not see the vortex form. One moment was confusion and pain, the next he was floating in the grey nothingness that filled the place between worlds. All pain ceased. The yellow flash spot was gone, replaced by never-ending grey. The ringing in his ears stopped. The tranquillity of the passage cocooned him in a bubble of peace.

For the first time, Elian did not even feel the wrench as he entered the place between worlds. He would never have believed it possible to feel relief at being there, but the sudden cessation of pain and confusion was a blissful release. He had just enough time to think that to fall into such a peaceful place would not be such a bad thing before being hurled back into reality.

Lashing rain whipped him with stinging lines of icy cold as they emerged. The flash spot blossomed instantly before his eyes and his right ear felt as if someone had pierced it deep inside with a long needle. Within seconds he was not just flying blind and deaf but also soaked to the skin and freezing cold. The realisation that he might be about to die was a shock. From the moment he had left home he had faced dangers. Dragonhunters, extreme cold, demons and night dragons – any one could have killed him, but he had escaped each challenge unscathed. For the first time, Elian suddenly became painfully aware of the fragile nature of life. He could not tell the extent of his own injuries and his dragon could be dying for all he knew. Was it all about to end? Was he about to crash and die? Robbed of his sight and hearing, all Elian could do was to hang onto the pommel and hope that Aurora still possessed the strength to get them to safety.

Blurred shadows loomed around him. It felt as though they were amongst mountains again. Aurora bucked and rolled beneath him like a wild stallion. He knew in his heart that it was turbulence, but for an insane moment he thought she might be trying to throw him from her back. Sheer desperation kept his fingers wrapped tightly around the pommel. He squeezed it so hard that his fingers seemed to lock and he doubted he could have let go again if he had wanted to.

All at once Aurora gave a huge lurch, like a horse leaping over a high hedge, and the pummelling rain cut off in an instant. The wind that had tugged at his clothing and hair also stopped abruptly and Aurora's motion switched from a flying rhythm to a running one. Relief warmed him. They had landed.

Aurora slowed to a stop and Elian leaned forwards slowly until his cheek rested against her back ridge just in front of the saddle. His fingers would not let go of the pommel, so he bent over them. He had thought that when the rain stopped his vision would clear but, although the flash spot was receding, his sight remained blurred and no amount of blinking made any difference. He could no longer feel his right foot and he felt sick with the pain in his ears and eyes.

The lullaby of unconsciousness began to sing its drowsy invitation but Elian refused to listen to its dreamy voice. There was something important that he had to do. He could not afford to rest. Another voice, distant, harsh and demanding began to intrude. Was that Aurora? No. It was Kira! He could hear her. He wasn't deaf. He could hear Kira. She wanted something. *Legs bowed? Letter owed?* What was she saying?

He felt hands grip him on either side of his jacket and pull him to his left. *Let go!* That was it. She wanted him to let go. He wanted to comply. He trusted Kira but his fingers refused to obey. The hands searched under his body and prised his fingers free. A spike of pain shot down his right leg as he was eased out of the saddle. He gasped with the shock of it. More hands supported him from below, lowering him gently down Aurora's side. *Pell? Nolita?* Had they returned to Orupee? No, that couldn't be it. The weather had been fine in Orupee when they left. It was cold and raining here. Where was he?

Elian groaned. He was being carried, but by whom, or to where, he had no idea. Voices spoke to him – distant dreamlike voices. He tried to concentrate on what they were saying. They sounded both concerned and encouraging.

Suddenly he realised he was no longer moving. His bearers had put him down. Where had they taken him? When he tried to remember how long he had been carried for, he could find no frame of reference. It had felt like an eternity, yet it might have been a few heartbeats. He could feel a soft mattress under his back. Someone was bathing his right cheek and ear. A burning sensation was spreading across his face that helped him to focus. He opened his mouth to speak, but no words would form on his lips.

Excruciating pain suddenly roared in his right thigh, causing him to buck violently on the mattress. He cried out, feeling the vibrations of that cry rattle around inside his head. For some time afterwards his teeth ached and his ears felt as though they were bleeding. Gentle pressure was being applied to the leg. It hurt, but nothing like that first shock of pain. This felt comforting. Blurry shapes moved around him, bent over him and made faint encouraging crooning noises.

'Rest, Elian.'

'Aurora?' he asked. His single word question asked many things – how are you? Where have you been? Where are you now?

'I'm all right,' she said wearily. 'But you are injured. Rest. We're at the enclave of the dawn dragons.'

'Why couldn't I reach you? I couldn't hear your voice.'

'I'll try to explain it when you've had some rest, dragonrider,' she assured him. 'Now close your eyes and sleep.'

She was gone again. He felt her presence cut off as if a door had been shut between them. What was going on? He desperately wanted to know, but now that he had heard from her he could feel his body relaxing. The dreamy lassitude he had felt after they landed began to sing to him again. This time he did not resist. Within seconds his body began to feel lighter and he felt himself drift loose on the dark tide of sleep.

Chapter Seven
Pursuit

Nolita awoke with a start. The *whooshing* sound of retreating dragon wings brought her to her hands and knees in an instant. She was alone in the shelter. She crawled outside. There was no sign of anyone around the campsite. Had they all left her? The tightening sensation of panic gripped her chest making it difficult to breathe. She could feel her heart rate accelerating as fear swept through her.

Fumbling for her boots, she struggled into them and stumbled across to the edge of the trees. She arrived just in time to see Aurora disappearing into a swirling grey vortex at the far end of the lake. The vortex collapsed behind her. She was gone. But who had gone through ahead of her? Had anyone stayed behind? The thought of reaching out to Firestorm made her chest tighten further. She interlocked her fingers and squeezed them together until sharp spikes of pain forced her to ease off the pressure.

'*Shadow and Pell are still here.*'

Firestorm's voice in her mind made her jump and then shudder. She doubted she would ever get used to the feeling of having a voice talk to her inside her head. She hated the sensation. It felt wrong – an invasion of her privacy. It left her nowhere to hide. She knew the dragon could hear her thoughts, but she did not know how much he could hear. Were there any corners of her mind that he could not reach?

'*Pell is behind you and to your right,*' Firestorm continued. '*He's watching you from the trees. I cannot say for certain, but I think he is taking pleasure from watching your discomfort.*'

'*Thank you,*' she thought back, projecting the words at the tenuous bridge she knew to be the path between their minds.

She did not need to look for Firestorm. She could feel where he was. Since she had accepted her role as a dragonrider, she had gradually become aware that she had gained another sense relating to her dragon. Even when she could not physically see him, she could feel where he was. At the moment she sensed he was tucked away among the trees a little way down towards the lake.

It would be a useful sense if I could extend it to feel the presence of others, she thought. *Then I would have known that Pell was still here.* Anger slowly began to replace the feeling of panic. The muscles across her chest relaxed and she drew in a deep breath. Pell knew her well enough to understand what she would be feeling at the moment. He had been cruel to deliberately conceal himself. Was this what the trip to the day dragon enclave was going to be like? Was Pell going to deliberately provoke her panic attacks to satisfy some twisted form of personal entertainment?

The fire of her anger burned hot in her gut, but she controlled it. If Pell drew pleasure from watching her panic, how much more would he gain from seeing her get angry? Nolita had no desire to further his pleasure. She bottled her anger up deep inside. He would want to go soon and there were things she wanted to do before they left.

Making her way back to the campsite, Nolita collected her water bottle and her precious bar of soap. Pell walked into the campsite as she was leaving for the lake. She ignored him. She needed to wash. Washing would help calm her.

'Don't be long,' he called out to her as she left. 'We need to get going soon.'

'I'll take as long as I want,' she muttered under her breath.

The water of the lake was cool and refreshing. She washed her hands first. Four turns of the soap for the first wash, three for the second. Her face and hair were next. The cool water felt wonderful as she soaped her hair. When she was finished, she wrapped the soap in its special cloth, wrung as much water from her hair as she could and walked back up to the campsite.

Pell was packed and ready to go. He had revived the fire and was sitting next to his saddlebags chewing on a piece of roasted meat.

'Feeling better?' he asked. 'Throw your things in your bag. We'd better get south of the mountains as early as possible. We don't want to draw unwanted attention to ourselves.'

Nolita did not trust herself to give a civil answer, but she went straight to the shelter and pulled her saddlebags and her blanket out in preparation to pack. The process of packing was another of her calming rituals. She spread the blanket on the ground outside the shelter and quickly took everything out of the saddlebags, laying them with precision in a neat pattern across the blanket. Once everything was laid out, she rapidly repacked, slotting and folding with careful accuracy to make the best use of the space inside the bags. The routine served a double purpose. It ensured that she never inadvertently left

76

anything behind and the repetition helped to calm her nerves and prepare her for the day.

Pell gave a pointed sigh as she began, but then watched her in silence. The process did not take long. Nolita had it down to a fine art.

'Are you going to eat before we go?'

'No,' she said, slinging her saddlebags over her shoulder. 'I'll grab something later.' She set off through the trees towards where she could feel Firestorm was waiting for her.

'Your choice,' Pell muttered with a shrug. He stamped out the little fire and carefully emptied his spare water bottle over the ashes to cool any remnant hotspots. With one final glance around the campsite to make sure they had not left anything behind, he headed off towards Shadow. The ferns that had covered the roof would begin to rot quite quickly, but the framework of the shelter would last a lot longer. It was an expected courtesy to leave such structures for other travellers.

'*Is everything all right, Pell?*' Shadow asked as he approached.

'*I've felt better,*' he admitted. The sight of the sleek lines of her gleaming scales between the trees lifted his mood. He could not help feeling good when he was close to her. '*I'm just a bit irritated that we have to look after Nolita and Firestorm while*

Elian and Kira get to go off adventuring, that's all.'

'We're going to the day dragon enclave,' Shadow said. 'Not many night dragons have ever gone there. We will be treading a path of adventure unseen by the likes of Segun.'

'Trust you to see the positive!' Pell said, smiling in spite of himself. 'You're right, of course, but I'd be happier if we were going alone.'

'I don't think going without them would be a good idea,' Shadow observed. 'The day dragon enclave will, doubtless, be pleased to see their newest heroes. I'm not sure we would be welcome without them at our side.'

'I give up!' Pell laughed aloud. 'Your logic defeats me, Shadow. I'll try to remember not to grouch in your presence any more. Let's go.'

He reached her side, leaped nimbly up her foreleg and slung the saddlebags into position. It took a moment to buckle them into place before he mounted the saddle and slotted his feet into the stirrups. Shadow moved out from between the trees. Firestorm and Nolita were already waiting. The two dragons looked at one another for a moment before leaping forwards and into the air.

The dragons climbed, side by side, into the bright early morning sky, angling first across the sparkling water of the lake and then, as they gained height, turning southwest across the hills and flying towards the western mountains of central Orupee. They could not avoid crossing the range, but prudence dictated that they avoid the immediate area around the Oracle's cave. The range curved around to the southwest here, so they could parallel the mountains for a long way before they would be forced in among the peaks.

Pell glanced across at Nolita. Her complexion was usually pale but, from this distance, her face looked positively milky-white and her blonde hair streamed out behind her like a golden banner. At first he thought it was the distance that made it appear as if her eyes were closed. The sun was bright and low. When combined with the wind rush it was reasonable to slit one's eyes, but there was something unseeing about the way she was holding her head. Was she really so scared that she could not look? Pell closed his eyes for a moment to see what it was like to fly blind. He did not keep them closed for long.

Firestorm appeared to be working hard to climb at this rate, but Pell could feel that Shadow was virtually cruising. Pride for his dragon swelled within him. She was an incredible dragon – strong and powerful. No matter what Segun and the other riders from the night dragon enclave did, they could not change her. Shadow was unique – and she loved him. For all his dreams of power and status, when it came down to the core of what life was all about, she was all that mattered. Since meeting her, he had felt complete, as if a part of himself that he had never realised was missing had clicked into place.

His feelings for Shadow made it doubly difficult to understand Nolita's continuing fear. How could she not feel the same way about her dragon as he did about Shadow? How could she remain so fearful when her bond with Firestorm allowed her to reach inside him and touch the very depths of his love for her? It did not make any sense. Unless her bond with Firestorm was not like the one he enjoyed with his dragon. Wasn't every bond between rider and dragon the same?

Pell had never really stopped to wonder much about the other riders and dragons before. He had been so caught up in his own plans and goals that it had not crossed his mind to think through such questions. With Kira around, he had felt compelled to focus on controlling the group. Now that she had gone with Elian, he had lost his chief source of irritation and competition. Nolita would go where he led.

'It looks as if trouble is set to plague us,' Shadow warned, interrupting his thoughts.

'What sort of trouble?' Pell asked, instantly alert and scanning the sky around them.

'Three night dragons approaching from behind and to the left,' she said, her voice in his mind sounding grim. 'The leader is called Darkhorn. He is telling me to land. They are angling to intercept. What do you want me to do?'

Pell looked left to locate the night dragons, and then ahead and left to the mountains. The range was still a long way off and they were closing towards the peaks at a shallow angle. He could not see the night dragons yet, but he did not doubt Shadow's word. If she was right and they were already positioning to intercept, then any thoughts of hiding were out of the question. It was run, fight or surrender. Pell had no intentions of surrendering, but the other two options did not seem likely to meet with success.

'Are Firestorm and Nolita aware?' he asked.

'Yes.'

'What do your instincts tell you, Shadow? Should we run or fight?'

'I think we should try to run,' she said. 'Firestorm is brave, but the odds do not favour us in a fight. The three night dragons are still some way off. If we run it will take them some time to catch up. They may give up the chase if they think it is going to be too much effort. I imagine their persistence will depend on their orders.'

'That makes sense but which way should we run?' he asked. 'If we turn and make a run for the mountains, we create a bigger corner for them to cut.'

'You are right,' she replied. 'That would not be a good idea. Let us make them run as far as possible by keeping them directly behind us.'

Shadow eased into a gentle turn to the right, adjusting her heading by about twenty degrees before rolling back into straight and level flight. She also began to accelerate. Pell looked across to see if Fire-storm had followed. He had. The blue day dragon's wings were pumping hard to keep up with Shadow, but there was tightness in the set of his face that conveyed determination. Pell was surprised. He had never noticed facial expressions on dragons before, and had not thought them capable of demonstrating emotion in that way.

Pell's heart raced as Shadow powered westwards, but it did not take long for the initial surge of adrenalin to die away. Craning his neck, he scanned the sky for any sign of the pursuing dragons, but he could see nothing. It was hard to force his eyes to focus much beyond Shadow's tail, which rose and fell in mesmerising waves behind them.

Looking ahead, he could see wispy puffs of fair-weather cloud beginning to form slightly above them in the otherwise cloudless sky, but Pell judged they would not grow into anything substantial enough to offer cover. Shadow had climbed to about a thousand spans. She was no longer climbing, instead concentrating on setting a pace that was fast, but one that she and Firestorm could maintain for a long time if necessary.

As time passed, the feeling of danger faded. Aside from the slight increase of the wind rush and higher tempo of the dragons' wingbeats, the flight felt much like any other. The countryside slipped past beneath them. The sun was climbing fast, its bright rays making it a pleasure to be airborne.

'*How are we doing?*' he asked Shadow eventually.

'*They are no longer gaining,*' she replied. '*I hoped they would give up, but they are still following. Darkhorn is livid that we have not heeded her warning. She has made it clear that they will not give up the chase. If you look directly behind us and at a similar level, you should be able to see them.*'

Pell twisted again and looked for some time. He was about to give up when he finally spotted them. Three tiny black dots, unmoving. Were they really at a similar level? They looked as if they were lower in the sky, but Pell knew that perspective could play tricks on a person's eyes in the air. He did not doubt Shadow's word. If she said they were at a similar height, then they were.

'So what now?' Pell asked. 'Do you think they're bluffing? Will they give up if we run fast enough?'

'It doesn't look that way,' Shadow admitted. 'It looks as if we are in this race for the long haul.'

'What about Firestorm? Can he keep up this pace?'

'Firestorm is already proving his resilience,' Shadow said, a note of surprise and grudging respect colouring her tone. 'It must be hurting him to fly at this speed, but he has not asked me to slow down and he is showing no signs of fading yet. Only time will tell how long he can keep going.'

Chapter Eight
Dawn Dragon Enclave

Elian woke suddenly. His body felt slick with sweat and his heart was beating hard. There was a painful ringing sound inside his head and everything around him was blurred. He blinked several times. The surface of his eyes stung with each movement of his eyelids and his clarity of vision did not improve. He tried to sit up but, as soon as he moved, it felt as if the room had been picked up by a giant and tossed, tumbling into space. The movement also triggered a spike of pain in his right leg.

He flopped back down against the soft pillow with a groan and closed his eyes. Even with his eyes closed the tumbling sensation remained. It made him feel nauseous. What was happening? Was he sick?

Memories of the short trip through France began to flood back. The explosion. The pain. He was injured. Was it bad? Was he blind? Deaf? A surge of panic began to build inside him.

'Calm down, Elian. I am here.'

'Aurora! What happened? The world is spinning out of control. I can't see anything.'

'Relax,' she soothed. 'You are safe now. One of the flash-bang weapons in the other world exploded right next to us. You suffered a nasty wound to your right leg. I can feel your disorientation. I do not understand the sensations you are experiencing. I will get Fang to send Kira to you now.'

'Your voice . . .' Elian began. He paused for several heartbeats before finishing. '. . . it sounds so weak. Are you all right?'

'It is my own fault,' she admitted. 'When the flash-bang struck, I panicked. I tried to form a gateway before true dawn. I succeeded . . . just. But it drained me terribly. It was all I could do to fly the short distance to land here in the dawn dragon enclave. I will need to recuperate for some time before I fly again.'

'I don't think I'll be ready to fly for some time, either.' The spinning sensation was gradually slowing. Elian tried opening his eyes, but he was quick to close them again. 'I can't focus and my head is full of noise.'

'Neema and Shimmer have gone to fetch a village medicine man they think might be able to help,' Aurora

86

said, her voice surprisingly positive. 'He has a reputation for working miracles.'

'Neema?'

'Neema is a dawn dragonrider here at the enclave,' she explained. 'Her partner, Shimmer, is a particularly fine-looking dragon. His scales are most attractive.'

'Reeeally?' Elian said, intrigued by her description. 'Attractive! That's the first time I've heard you describe another dragon in that way.'

'Well,' she replied, giving no hint that she had picked up on his inflection. 'That is because Shimmer is more attractive than most. When your sight has recovered, you will see what I mean. He has always been popular as a mate with female dragons from the day, dusk and dawn enclaves. He has recently flown Blaze. She is in the hatching cave brooding over her clutch of eggs.'

'Flown Blaze? What do you mean . . .?' Elian began to ask. He stopped as his mind connected the phrase with the clutch of eggs. 'No, never mind. I get the idea, but you say Shimmer is popular as a mate. Don't dragons have life partners?'

'No, Elian,' she said. 'Our mating traditions are very different from yours. We live for a long time, so nature has restricted our urge to mate. Dragons spend much of their lives travelling alone. When the urge does come, we may be a long way from the mate of our choice.

Finding a suitable mate then becomes a matter of—'

'Enough information!' Elian interrupted, feeling his cheeks flush. '*I think I get the idea.*' This conversation was straying into areas he was not comfortable with, and he got the distinct feeling that Aurora was willing to explain things in more detail than he was ready for.

'*I do not understand your embarrassment, Elian,*' Aurora said. '*Mating is a perfectly natural course of events for every creature, man or dragon. It is nothing to be embarrassed about.*'

'No. I know,' he replied, mentally squirming as he tried to think of a way of diverting her. '*So have you ever . . .*' He left the sentence hanging, mentally kicking himself for starting down this path.

'*Mated?*' she finished, her tone still matter-of-fact. '*No. The urge has never touched me. It is not unusual amongst dragons for a female not to mate. When we do, we tend to produce a lot of offspring. Blaze has over forty eggs hardening in the hatching cave. Perhaps a third of those that hatch will survive. Life as a young dragonet is hard. But if we mated with the frequency you humans did, the world would be overrun with dragons very quickly.*'

Elian could appreciate that. He tried to work out how fast the dragon population would expand if dragons produced offspring more regularly, and the numbers got very big, very quickly. It was a useful diversion to take his mind from the pain in his leg and his eyes.

A sound broke through the white noise that was blanketing his hearing. Someone had entered the room and was trying to talk to him. It took a moment for him to recognise Kira's voice. He tried opening his eyes again. The room was still blurred, but he could see moving shapes. Kira's voice became clearer as she approached to his left.

'Elian? Can you hear me?'

'Yes I can hear you,' he said, his own voice sounding strange inside his head. 'But only through my left ear, and that feels strange. There is a constant sort of crackly, ringing noise that's making it hard for me to hear, and I can't see properly. My vision is constantly out of focus.'

'I've brought someone to see you,' she said, speaking slowly. 'He's come to help make you better.'

The blurry outlines switched around and another person leaned over him. Cool, strong hands touched Elian's cheeks, the fingers gently feeling around to the back of his neck, carefully avoiding his ears.

'Hello Elian.' The deep male voice was clear despite the noise inside Elian's head. 'My name is Haithan.'

Although he was out of focus, Elian could make

out that Haithan had an impressive flowing mane of golden hair. The man's hands made a complete sweep of Elian's upper body, gently feeling for anything unusual. Elian tensed as the hands reached the tops of his legs.

'It's all right,' Haithan assured him, his deep voice almost hypnotically calming. 'I won't touch the area around the obvious wound. I just want to make sure there's nothing else that your friends have missed before I begin working on what we can see.'

'Thanks.' Elian relaxed slightly, but a certain amount of tension remained in his limbs, shoulders and back until Haithan had finished his investigation.

'Some sort of exploding weapon, I'm told,' Haithan said, his deep voice thoughtful.

'Yes.'

'Have you tried to move since you woke up?'

'I tried to sit up a little while ago, but I felt dizzy and was nearly sick,' Elian admitted, feeling embarrassed by his weakness. 'Also, my sight is blurred and my eyes hurt whenever I blink. There's a buzzing ring in my ears that won't stop, and my body feels as if it's been side-swiped by a dragon's tail. Aside from that I feel ready to wrestle a lion.'

'Hmm!'

'What do you think, Haithan?' Kira asked. 'Can you help him?'

'Oh, yes,' he replied. 'I can help. He's a very lucky young man.'

'Excuse me if I think differently,' Elian muttered.

'What I mean is, you're lucky to be alive,' Haithan said. 'The weapon, or at least a part of it, is lodged deep inside your thigh. If it had hit you in the chest, it is unlikely we would be talking. The surfaces of your eyes have been slightly cooked by the heat and flash of light. That damage will likely heal on its own in a few days. The fluid that your eyes produce should work its own miracle there. Your hearing and dizziness is more worrying. There is something I can try that might help stop the disorientation. Which ear is the worst?'

'The right,' Elian said, a lump forming in his throat. He tried to imagine what his leg looked like with the weapon stuck in it. Why hadn't Kira taken it out? What could Haithan do?

'All right,' Haithan said. 'Relax and move with my hands. Good. Now, I want you to count slowly with me to twenty. We're going to do it three times with your head in different positions, all right?'

'I'll do my best.'

'Good. Kira and Neema, I'll need your help as well,' Haithan continued. 'Soon I'd like you to help me roll Elian on to his left side. Be careful not to touch the wound on his leg. I want to avoid putting any pressure on it.'

'Where would you like us to stand?' Kira asked.

'Please stand to his right. Now, Elian, are you ready to begin?'

'Yes.'

Haithan moved to stand next to Elian's head. 'Here we go then.'

The healer cupped his hands over Elian's head and tipped it about forty-five degrees to the right. He began counting. 'One ... two ...' Elian joined in softly. '... nineteen ... twenty.' The hands tilted Elian's head through ninety degrees to the left. The counting started again. What good did Haithan think this would do? Elian couldn't begin to imagine, but he played along. '... eighteen ... nineteen ... twenty.'

'Roll him now,' Haithan ordered.

Two pairs of hands gently lifted Elian's body until he was lying on his left side. Haithan's hands turned Elian's head further to the left until he was looking straight down at the bed. 'One ... two ...'

Elian could not have moved his head even if he had wanted to. The hands that held him were strong and firm. He counted along with the deep voice until they reached twenty again.

'All right. He can sit up now,' Haithan said.

92

Hands began to tip Elian upright. He tensed his stomach ready for the whirling disorientation he had suffered before. To his amazement, there was none. No spinning. No nausea.

'That's amazing!' he gasped. 'What did you do? My head isn't spinning at all. My balance feels perfectly normal!'

'It's a trick I learned some years ago,' Haithan said, his deep voice containing notes of pleasure at his success. 'I wasn't sure it would work on your injury, but the fact that it did furthers the proof of my theory. One of the hunters in my village used to suffer vertigo. I tried lots of different things to cure him and in the end it was this simple manipulation of his head that worked. Certain types of dizziness are due to something becoming misaligned inside the ear. Balance is controlled there. By trial and error, I found that this simple manoeuvre of the head can restore that alignment.'

'It seems the rumours about you are justified,' Kira said. 'You can work miracles.'

Haithan laughed. His laughter was as rich and rolling as his speech. 'Miracles! Some call my work miraculous. Others accuse me of magic. I don't really care what people say. It brings me pleasure to help others.'

The ringing in Elian's ears had not abated, but he

could follow what was being said. The voices were distant, though he knew the people speaking were right next to him.

'Can you do anything to stop the ringing in my ears?' he asked. 'I can't hear anything in my right ear and there is a sharp pain there unlike the more general pain I'm feeling elsewhere.'

'I can prepare some drops for your ears,' Haithan replied. 'There is a plant that grows in the rain-forests nearby with properties that should help. The solution doesn't keep long, but it isn't difficult to make. I suspect you've damaged the membrane inside your right ear. It should heal in time and you will regain a measure of your hearing. A day dragon's healing breath might restore it, but I don't know for sure and we don't have one to hand.'

'There wasn't a lot of certainty in any of that,' Elian muttered.

'No,' Haithan agreed. 'People do not all heal in the same way, I've observed. Why do two people with the same symptoms react differently when given identical treatment? It is a mystery that has eluded me so far. I can only help you by using methods that have worked for the majority of people. If you have the will to get better, then this often tips the balance.'

'I thought everyone who was ill wanted to get

94

better,' Kira said, clearly unimpressed by Haithan's explanation.

'It's surprising, I know,' he said solemnly. 'But I've seen some people make a full recovery from horrific injuries who I thought would not survive a day, and watched others, who were really not very ill, fade away and die for no reason. The desire to live appears to be the most powerful healer of all.'

'What can you do for his leg?' Kira asked.

'That is a tricky question,' Haithan replied.

Elian watched the blurry figure bend over to look more closely at his right thigh. He winced in anticipation of pain, but Haithan did not touch him. Instead he studied the wound for some time before straightening.

'I think there is only one answer,' he said, his voice slow and thoughtful. 'I'm going to have to cut out the remains of the weapon. If I leave it inside, the wound could turn bad and cause the flesh to rot. I've seen a far lesser wound flood a man's body with poison and kill him. I can make a solution that will temporarily take all feeling from the area while I take the thing out, and another to clean the wound before I close it back up again. It should not take long.'

'What about Aurora?' Elian asked.

'What about her?'

'Can she not try to heal me?'

'Do dawn dragons breathe the healing fire?' Haithan asked. 'I did not realise.'

'No, they don't breathe it,' Elian explained. 'But they have it within them. Aurora healed Shadow, our friend's night dragon recently.'

'Is this true?' said another voice. 'How did she do that?'

The voice was female, but not Kira. It had to be Neema, Elian decided.

'I was not there when she did it, but the results of her healing were spectacular,' Elian said excitedly.

'I would heal you if I could,' Aurora said. *'But I have no energy. I could ask Shimmer if he would be willing to try, though. I can explain what he has to do.'*

'Even if this sort of healing is possible, I still think I should remove the weapon first,' Haithan suggested. 'You do not want to live out your life with that thing inside your leg. There's no telling what long term damage it might do.'

A cold clamp of fear squeezed inside Elian's chest. He did not want Haithan to cut his leg, but he knew in his heart that it was the right thing to do.

'Haithan is a good man,' Aurora said. *'His thoughts are positive and focused on your well-being.'*

Elian took a deep breath.

'My dragon tells me to trust you,' he said. 'If my dragon believes you worthy of trust, who am I to argue? I won't lie – I'm not happy about the thought of you slicing my leg further, but I trust Aurora. Do what you have to do.'

Chapter Nine
If Only …

'It's no good, Pell,' Shadow announced. 'Firestorm is struggling. He's done amazingly well to fly this fast for so long, but he tells me he can't keep up this pace any more. We're going to have to do something to throw Darkhorn and the other night dragons off our tail.'

Pell instinctively looked across at the blue dragon labouring along beside them. Judging by the position of the sun, they had been flying for over three hours. He could feel how tired Shadow was from flying at this speed for so long, so it was hard to imagine what it had cost the relatively small day dragon.

'Do you have any suggestions?' he asked.

'We could turn and fight, leaving Firestorm to go on ahead.' Shadow was trying to sound upbeat about the idea, but Pell knew her well enough to pick up on the subtle inflections in her mental tone. It was an option, but there was little chance of success.

'Can't Fire turn with us and help fight?' he asked. 'Three against two would give us better odds.'

'Fire is so tired his wingbeats are almost involuntary reflexes now,' she replied. 'He's worried he won't be able to manoeuvre well, and he doesn't think he will be able to raise much of a flame.'

Pell let his focus slip past Firestorm to settle on the range of mountains beyond. Shadow had incrementally edged their direction around until they were easing ever closer to the great line of peaks. A glance over his shoulder showed Pell that the three black dots had closed very slightly since he last looked.

'They're closing,' he noted. 'We need to act soon if we're going to lose them. What about going into the mountains?'

'I don't know this part of the mountain range very well,' Shadow admitted. 'But your thinking is sound. Our chances of getting away will improve if we can get amongst the peaks. I am going to lead Nolita and Fire into the next pass. Trying to outrun Segun's men in the open air has not worked. We will have more chance of losing them in the valleys. If we do have to fight, we would do well to limit our opponents' room to manoeuvre. Also, if I fly cleverly, we might be able to draw all of them after us, giving Firestorm and Nolita a chance to get away.'

Pell sensed Shadow communicating her intention to Firestorm, though he did not hear the words. They needed a valley between peaks that was deep enough that they could enter it without needing to climb. He could feel Shadow's fatigue and, although he felt sure she could manage a reasonable climb rate, he doubted that Firestorm would find the energy to climb at more than the shallowest of gradients.

There was a promising valley ahead. The opening was after the next peak but one. Shadow turned left, cutting in at an angle that took them very close to the mountainside. At the same time, she dropped down and behind Firestorm, allowing the day dragon to lead the way at his best pace.

Pell saw the chasing dragons react. They turned instantly to cut the slight corner. It appeared they were closing more quickly now, though Pell knew this to be an optical illusion. The closer they came, the faster their approach would seem. He had seen this before. He knew that as they closed the final few hundred paces, they would blossom quickly from their current toy-like, harmless appearance, transforming into an enormous and fearsome reality. It would not be long now.

A burning rush began to build deep inside Pell's gut. The familiar heat of adrenalin sharpened his senses and set his mind racing. There had to be something they could use to gain an edge. Anything that gave them an element of surprise might make all the difference, but Pell's only thought was that Shadow was the strongest dragon he had ever seen, with the possible exception of Widewing. Even though they were gaining, Pell could not imagine the three chasing dragons were any less tired than Shadow. In a straight power confrontation, one-on-one, Pell felt confident that Shadow could best any of the other dragons. But how could they twist the situation to give Shadow a chance against all three?

The green mountainside to their left reared ever closer and the sensation of speed began to build as they edged towards the slope. Texture and detail began to leap out at him as Fire and Shadow hugged the steep contours. Splodges of grey stone suddenly resolved into three dimensions as rocks appeared to grow out from the slope. Flashes of red and yellow flowers, invisible at height, streaked past as every nook and crevice became visible.

Pell loved the feeling of speed as he flew close to the ground. He was racing around the mountainside faster than anyone could ever traverse it on foot, or even on horseback. Riding a horse at a full gallop was exhilarating, but nothing could compare with skimming across the land on a dragon's back.

As they tipped into the turn around the mountainside and into the valley beyond, it felt as if Shadow's wing might catch on the ground at any moment. Looking ahead at Firestorm's relative height gave a better clue as to their real clearance. Even given Shadow's greater wingspan, they were maintaining a healthy margin of safety. At this speed the ground felt closer than it really was.

Another glance over his shoulder revealed the nearest pursuing night dragon had closed to little more than twenty dragon lengths. The other two were quite some distance further back, but there was not enough distance between the front-runner and the other two dragons to isolate it. The other two would catch up quickly if they were to turn and fight.

He looked ahead. The valley had steep sides, but its base was relatively broad. A seed of an idea began to germinate, but it would not work here. He needed somewhere tighter.

'*Start looking for a narrow canyon, Shadow,*' he told her. '*I've just thought of something we can do to make life very difficult for the dragons behind us.*'

'*You're going to have to hang on really tight, Nolita,*' Fire warned, the strain of the chase evident in his voice. '*The night dragons are stronger than I am. I've given it my best, but I can't stay ahead of them any more.*

They are catching up fast. Things are likely to get rough.'

The familiar black wall of fear began to rear inside Nolita's mind as she absorbed her dragon's words. Even though they had been chased for hours, this flight had felt no more frightening than any other – until now. Fire had been flying faster than normal, but Nolita had not seen their pursuers in the past three hours, and their threat had seemed distant and insubstantial. Suddenly she felt tiny and vulnerable. What could she do if the dragons began to fight? They were so big and she was so inconsequential.

She could feel Fire's fatigue in every fibre of her body. He had thrown all his energy into running from the night dragons. Now he was exhausted. What could he hope to do against three bigger, stronger dragons? Shadow would be a powerful ally, but even she could not hope to tackle three at once.

'What can I do to help?' she asked, feeling more comfortable with speaking the words aloud.

'Unhook your feet from the stirrups,' Fire told her. *'Pull the stirrups up onto the saddle and loop your hands through them. Stay low to my back. Twist the stirrup straps around your wrists so that even if you slip, you will not fall. I need to concentrate on keeping the night dragons at bay and I might have to make sudden changes*

103

of direction to keep them guessing. I don't want to have to warn you of each turn. It will slow us down and make us more vulnerable.'

'I'll do my best,' she promised.

'I know you will,' Fire assured her. 'And I'll do mine. Between us we'll be fine.'

The words were brave, but she could feel no conviction in them. The more she got to know Firestorm, the more human some of his actions were. He was putting a brave face on a bad situation, just as her mother used to do.

She slipped her right foot from its stirrup first. Having done so, she could not see how she was going to reach the leather strap that attached it to the saddle. It meant stretching quite a way down Fire's side and she was not sure she could bring herself to lean that far.

'If you don't do it now while I'm in steady flight, you won't be able to do it at all,' Fire told her, following her thought processes. 'You can do it. Hold onto the pommel with your left hand and reach down with your right. It will be fine. Trust me.'

To Nolita's surprise she found that she did trust Firestorm. She did not like looking down when they were flying. Her head tended to spin and she felt sick when she thought about how high they were. To reach the strap, she had to lean a long way down and she knew that closing her eyes would be a foolish thing to do.

Forcing herself to lean away from her relatively stable seated position, she gritted her teeth, gripped the pommel with her left hand and tried to reach down. Keeping her eyes focused on her goal, she inched her hand further and further down Fire's side. On her first attempt her fingertips almost reached the strap – almost, but not quite. After several seconds of creeping her fingertips down Fire's side, she hauled herself back upright in the saddle and took several deep, slow breaths to calm her pounding heart.

Nolita had no desire to try again, but she knew there was no choice. She had to do it. Twisting her body a bit to the right, she gripped the pommel even tighter this time and, flattening her chest against Fire's body, she stretched down. Her change of position was just enough. Hooking her middle finger under the strap, she drew the stirrup up until she could grip the metal footrest. As soon as she was upright again, she put her hand through the loop and twisted the leather around her wrist a couple of times.

Finding the confidence to lean down to the left was far easier. With her right hand secured in the right stirrup, Nolita was quick to secure the left. It felt strange to have her legs dangling loose, but surprisingly comforting to have her hands twisted into the leather loops. The stirrup straps were not long enough to allow her to put her hands back on the pommel, so she gripped the leading edge of the saddle instead.

'That feels a lot safer,' she told Fire. 'I would prefer to have the stirrups as well, though. It feels strange to have my feet dangling loose. When we next land somewhere close to a tanner, I'm going to get some leather loops sewn onto the pommel. I feel so much safer with my hands secured like this.'

'Good idea,' he replied, skimming around the mountainside and into the mouth of the wide valley beyond. 'You never know, you might start a trend.'

'Where's Shadow gone?' Nolita gasped, noticing for the first time that the enormous night dragon was no longer beside them. The mountainside was racing past to their left, but there was no sign of Shadow to their right and she had been leading the way ever since they had first detected the other night dragons.

'She's directly behind us, covering our backs,' Fire told her. 'Segun's men are getting very close. If my wings weren't so tired, I'd feel more confident of out-manoeuvring them. I know a few tricks that should hold them off for a while. Shadow is strong and will fight *bravely, but I would give a lot to have Aurora and Fang here now.'*

'So would I,' Nolita muttered. 'So would I.'

Elian thought he was dreaming at first, but there came a point when he knew he had crossed the boundary from sleep into a conscious awareness of his surroundings. The sensation was almost like floating in warm water. He opened his eyes. It took a moment to remember his injuries. The blurriness did not seem as bad as before. Colours appeared more discernable and he felt he could make out more detail of his surroundings.

'Are you awake? How are you feeling?'

It was Kira. He did not know how to answer. Nothing felt real. Everything was still fuzzy, but the ringing in his ears had receded to a bearable background buzz – annoying, but no longer a major problem. He could not decide if he had heard her voice in both ears, or just in the left.

'All right, I think,' he mumbled eventually. 'What time is it?'

'Early afternoon,' Kira replied. 'You've been asleep for more than a day.'

'No wonder I can't feel my back,' he said sleepily. 'I never oversleep. It must be something to do with the injury. I still feel groggy.'

Every word brought him more in touch with his surroundings. Kira was sitting on a chair beside his bed. He could not make out her expression, but there was a note of relief in her voice. How long had she been sitting there watching over him, he wondered?

'Haithan gave you a sleeping draft to keep you unconscious while he took the metal fragment from your leg,' she said. 'Do you remember anything of what he did?' Something in her voice told Elian that her respect for the healer had grown.

'Not a lot,' he admitted. 'Where is he now?'

'Neema and Shimmer took him back to his village. He said there was nothing more he could do for you at the moment, but Neema tells me that Shimmer has been talking to Aurora. He's keen to try using his inner fire to heal you further once you're able to move to him. Apparently both Shimmer and Blaze were amazed to learn they had powers of healing. The archives here do not mention this ability.'

'What about the others?' Elian asked.

'What others?'

'The other dawn dragons, of course.'

'There are no other dawn dragons, Elian,' Kira replied. 'Shimmer, Blaze and Aurora are the only three dawn dragons in Areth who have riders at the moment. We happened to catch them both at home because Blaze wanted to lay her clutch of eggs somewhere comfortable, but the enclave is often empty for years at a time.'

'Only three!' he exclaimed. 'I had no idea. I mean, I always knew that Ra was special, but I didn't realise dawn dragons were *that* rare. Any idea how many there are without riders?'

'No, I didn't think to ask.'

'What about the orb?' Elian asked, feeling suddenly excited and a lot more awake as he remembered why they had come to the enclave. He propped himself up on his elbows as he continued. 'Is it here? Have you found any clues?'

'No,' Kira sighed. 'Nothing. I've had a good look around the enclave, but there's no sign of it. The enclave is tiny by comparison to the homes of the day and night dragons. Neema and Blaze's rider, Tarl, don't know anything. Neither do their dragons. I don't think the orb is here.'

Elian thought hard for a moment. This was a blow. Things were not going as he had expected. Until now, the path of the quest had led them from one orb to the next, laying itself out before them in an obvious path. Despite their best efforts, though, time was running out. They could not traipse all over Areth looking for clues. He had expected to find something here that would at least give them a lead on where to look next.

'What about using the dusk orb to help with the search?' he asked.

'I tried.' There was a strange inflection in Kira's voice. She sounded distant, as if she were reliving the experience. 'I sent my vision into every nook and cranny in the caves here, but I found nothing. I even snuck in to look around the hatching chamber, but somehow Blaze knew I was there. She got very agitated and I didn't want to upset her or Tarl, so I didn't stay long. There was little point. The chamber was bare. Aside from Blaze and the eggs, there was nothing but stone – floors, walls and ceiling.'

'Secret passages, or chambers?' Elian mused.

'I'm not sure,' Kira admitted, 'but I don't think so. When I'm using the orb, my sense of vision is enhanced. I think I'd know if there was something there.'

Elian slumped back, disappointed. His mind was racing, but while he was unable to see properly there was little he could do. Much as he hated to admit it, he was stuck here for as long as it took to get better.

Chapter Ten
Dangerous Manoeuvres

'*Split now!*'

Shadow rolled hard to the right, simultaneously powering upwards as she veered away from Firestorm's tail. But Darkhorn had followed. She was really close now and Pell gasped as he saw the dragon's mouth open wide. With a sudden surge, the night dragon lunged forwards to bite Shadow's tail.

'SHADOW!' It was all the warning Pell could formulate.

Whether it was instinct on Shadow's part, or whether she had been aware of the intention of the other dragon all along, Pell did not know. One heartbeat her tail was gently undulating with the natural rhythm of flight, the next it whipped into a tight arc to the left, swiping at the lunging night dragon and giving it a mighty slap across the head.

Darkhorn screeched with anger. Her head reared slightly and she lunged again, only for Shadow to arc her tail the other way and deal her another slapping blow. The rhythm of Darkhorn's wingbeats seemed to falter momentarily and she dropped back a short distance. Pell felt a fierce burning in his stomach and his hands gripped tighter than ever on the front of the saddle. He could see that further back, the two remaining dragons had split, one following Shadow and the other tracking Firestorm.

'*Darkhorn and Longtail have decided to follow us. Deepshade is going after Firestorm,*' Shadow announced.

'*What will Firestorm do?*'

'*I don't know,*' she replied. '*But Firestorm and I agreed that splitting up is the best course of action.*'

They had given Nolita and Firestorm a bit of breathing space, but it seemed this was the best he and Shadow could do for now. Pell did not fancy Firestorm's chances against Deepshade, but there appeared little hope that he and Shadow could deal with Darkhorn and Longtail *and* catch up with Firestorm in time to assist. He was pleased the dragons had made the decision. Normally he liked to lead, but his promise to Elian that he would get Nolita to the day dragon enclave was fresh in his mind. It had felt like he was abandoning that promise less than a day after making it, so his guilt was eased by the dragons taking charge.

Shadow headed for the high valley to their right and Pell had an idea that rapidly developed into a plan. Shadow's choice of route was not suitable for the trap he had in mind, but Pell knew what to look for now. He could feel that Shadow was confident she was stronger than their two pursuers. She was aiming to stretch the two chasing night dragons beyond their limits. If she could, she would force them to give up the chase. If not, she hoped to weaken them enough to swing a fight in her favour. With the immediate odds down to two against one, things were beginning to feel slightly better and Pell was sure that with a little bit of luck, things could swing to their advantage.

'*Are you sure you can climb fast enough to make it over that pass?*' Pell asked, looking up at the tight 'V' shaped valley.

'*There's only one way to find out,*' Shadow replied, her voice set with determination. '*But I'll make it. Trust me. If I have to land and run the final bit, I'll still beat Darkhorn and Longtail to the other side.*'

Despite the steepness of the valley, Pell believed her. They had been flying at speed for hours, yet Shadow still had strength in reserve. So powerful were her downward strokes it felt as if she was stamping up a flight of stairs. The nearest of the night dragons, Darkhorn, was beginning to drop behind, but Longtail had been quick to respond to Shadow's tactic. She had started her climb from further back and did not have to climb at such an extreme rate.

Pell's head was beginning to spin as he kept looking forwards and back, forwards and back. It was hard to decide where the biggest danger threatened. The rocky pass ahead was narrow and steep. The two dragons behind looked angry and determined. The temperature was dropping as they climbed and soon the air would thin, making it more difficult to breathe. He knew Shadow was intending to descend again as soon as they were over the crest so he was not worried about breathing becoming impossible, but thinking was more difficult in this environment and he was keen to keep his mind alert.

The walls of the valley closed in on either side of them as Shadow hurled all her strength into the climb. The jagged rocks beneath gradually drew more and more of Pell's attention as they reached up towards him with teeth every bit as deadly as those of the attacking dragons. How Shadow thought she could effect a landing here, Pell did not know. Their crash landing on the ridge when the dragonhunters had chased them was vivid in his mind. Then he had been thrown clear into soft heather. There was nothing soft to land on in this valley.

'*We will not crash,*' Shadow assured him.

'*Glad to hear it.*'

Despite her assurance, they were getting closer to the rocks and Shadow was already putting her all into climbing as fast as she could. Pell tried to project their trajectory ahead to see where they were going to impact the rocky slope, but he could not quite see far enough. The valley twisted gently to the left ahead.

Tucked down in the narrow valley, they were no more than fifty spans clear of the rocks below and had about double that to either side. A night dragon screeched behind them. Pell wanted to look back, but was fixated on the ground ahead.

'*Darkhorn was forced to land,*' Shadow said, her voice sounding smug. '*She and her rider are unharmed, but she is not happy about having to climb the rocks to follow me.*'

'*How can any dragon land here?*' Pell asked, amazed to hear the dragon and rider managed to do so without injury.

'*Carefully,*' Shadow told him. '*It is not so hard for a healthy dragon, but it is easy to make mistakes when one is tired. Look! We're nearly at the top.*'

Shadow was right. They were just rounding the

slight turn and the top of the pass was no more than a few hundred paces ahead. They were only about fifteen spans above the rocks now, but the gradient was decreasing and Pell could see they were going to reach the crest without having to touch down. Secure in this knowledge, Pell glanced back over his shoulder. Only one night dragon was in sight. It was Longtail. Pell's eyes lingered on her for a moment. Her wings were beating hard and deep, but her labour was paying off. She would also make the top of the pass without having to land.

The change from climb to descent was abrupt. For the briefest instant, Pell felt lighter than air as Shadow coasted over the top of the pass and dipped into a steep dive. Another wide valley between lines of mountains was a welcome sight. He scanned left and right for any sign of a likely place to lose the night dragons.

'I think we'd do better to turn left along this valley when you can, Shadow,' he suggested. 'Take us deeper into the range. We should have more chance of losing them there.'

'I agree,' she replied.

Shadow initially followed the line of the pass on the other side of the ridge, giving several powerful accelerating beats to stretch her lead on the night dragons. She stayed close to the surface, skimming across the rocky terrain as she converted every span of height into forward momentum. Pell looked back and was surprised to see Longtail did not follow their lead. Instead she cruised out over the top of the ridge, maintaining her height, choosing to keep the advantages of visibility and potential energy rather than commit to a low-level chase.

As soon as Shadow realised what the other night dragon was doing, she made a deliberate turn to the left and eased away from the surface, converting some of her speed into height, but doing her best to fly fast enough to stay ahead.

'Longtail is wily,' she observed. 'I was hoping to draw her into a straight chase, but she's outmanoeuvred me and gained the advantage. I'm afraid we could be in trouble here.'

'I've got an idea that could tip the balance in our favour,' Pell said. 'We need to find a narrow box canyon. Preferably one with no room to turn around without landing.'

'What do you have in mind, Pell?' Shadow asked, curious.

Pell told her.

Shadow did not reply for some time. Pell could feel her thinking his suggestion through. When she did finally speak, her response was not quite what Pell expected.

'It is terribly risky,' she said slowly. 'If we're too close to Longtail when we try it, we could all get killed.'

'Surely everything's risky, isn't it? It's two against one. This gives us the element of surprise.'

'You're right,' she agreed. 'I doubt there are many dragons who would attempt it. My hesitation stems from my inability to decide if your plan demonstrates boldness or insanity!'

'There's a night dragon right behind your tail . . . and it's not Shadow!' Nolita gasped. With her body flat against Fire's back and her hands wound through the stirrups, the only way Nolita could look behind was by raising an arm slightly and peering underneath and back along the length of her dragon's body.

'This is where flying starts to get really interesting,' Fire replied. 'Hold yourself tight to my back and try not to be afraid. We're going to have to turn hard and often.'

'Where's Shadow gone?'

'She's drawn off the other two night dragons,' he replied. 'She and Pell are heading over the high pass to our right. The dragon behind us is called Deepshade. We're one against one now.'

Nolita was horrified. Pell had left her. She was alone with her dragon. That held more significance than the fact they were under attack. The familiar wall of black terror began to rise. Her breathing

118

quickened and she could feel her heartbeat accelerating. Her dragon's fatigue was making every sinew in her body ache in sympathy, yet from somewhere she could feel him summoning strength from a seemingly impossible reserve.

'*Turning right,*' Fire told her, simultaneously snapping into such a vicious turn, he appeared to be pivoting around his right wingtip.

A brutal force, far greater than any she had experienced before, crushed Nolita's body against the saddle. Nolita screamed. She could not help it. A deep inhalation swelled Fire's back beneath her. He arched his head on his long neck and blasted a stream of fire past her at the closing night dragon. She knew that he could breathe far hotter fire, but this blast seemed to cost him more than any she had felt before. Deepshade screeched, though Nolita could not tell if it was through pain, or rage.

'*Turning left.*'

The transition was so fast that her head spun with the speed of the rolling motion. For a brief moment, the force pressing her against Fire's back eased, only to return an instant later as he wheeled hard in the opposite direction. There was a slight bump, like a jolt of turbulence and the turn smoothed again.

'*That was close!*' Fang exclaimed. '*Deepshade turns faster than most night dragons. I've not got the energy to*

119

drag this out. We must take them down quickly, or they will kill us for sure. I'm going to try something I saw one of those flying machines do in the other world.'

Nolita was struggling to focus on what Fire was telling her. She did not know what had just happened, nor what he intended to do, and she was too scared to ask. Her fingers were gripping the leather of the stirrups so hard that she could feel her nails cutting into her palms. She screwed her eyes as tightly shut as she could and, stifling a scream, gritted her teeth together so hard that the muscles on either side of her jaw bulged fit to burst.

The rolling motion this time, as Fire reversed the turn, was not so severe. But instead of the crushing pressure decreasing as he rolled, it increased slightly and then very gradually decreased until it disappeared altogether. As the force decreased, so did the noise of the wind rush and Nolita began to feel a curious sensation, as if she was about to float free of her dragon's back. Still the steady rolling motion to the right continued. Was it never going to stop? What was Firestorm doing?

As gradually as the force holding her against Fire had eased, so it began to build again. Nolita no longer had any idea which way was up, but she did not dare to open her eyes. She could feel Fire brewing another jet of flame as his back swelled beneath her.

The pressure forcing her against the saddle continued to build, pressing harder and harder. Firestorm loosed another roar of fiery breath, drawing a second screech from Deepshade. This time, curiously, the sound originated ahead of them, and there was no mistaking it for anything other than a cry of pain.

A surge of triumph swept through the bond, almost overwhelming Nolita with its intensity. The emotion lasted barely an instant, only to be replaced by an equally potent wave of panic. Where the force holding her against her dragon's back had been increasing smoothly up until this moment, suddenly it peaked with savage power, squashing her flat against his back. As suddenly as it peaked, so the force was gone. With the release came a third flood of emotion – relief.

'It's over,' Fire told her wearily. 'Deepshade won't be following us any more.'

Nolita cautiously cracked open her eyes. They were flying very low along the valley basin. Glancing over her shoulder, she glimpsed the night dragon on the ground behind them. She saw movement. It was not dead.

'What did you do?' she gasped.

'I burned a large hole through her right wing,' Fire said gravely. 'It was a horrible thing to do, but I was left with no choice. Without a miracle, she will not fly again

121

for a long time. I am pleased she and her rider survived the crash. I would not like to be responsible for the death of another dragon, or worse, for turning one rogue.'

'But Deepshade was behind us,' Nolita said, trying to replay the sensations she had felt as they had manoeuvred. 'How did you turn the fight around so quickly?'

'I did something that I don't think any dragon has ever tried before,' Fire replied. 'I completed a full roll to the right.'

'A full roll? What do you mean?'

'I mean we turned to the right, but kept the roll element of the turn going until we turned through an upside down position and back to normal flight.'

'But that's impossible! I would have fallen . . . or at least been hanging by my arms from the stirrups,' Nolita insisted.

'If I had tried simply to roll, then yes, I imagine that is what would have happened,' Fire agreed. 'But I didn't. When we were in the other world I saw one of those flying machines complete a sort of looping roll. The machine was constantly pitching upwards relative to the man inside while rolling at the same time. It was the constant tipping upward movement that kept you from falling, though I'm not sure I did it quite right. I felt you almost leave my back as we got to the fully upside down position.'

Nolita was confused.

'I'm not sure I really understand what you mean. I still don't see how a roll could take you from in front of Deepshade to behind her,' Nolita said, slowly straightening herself in the saddle until she was sitting upright.

'Well, to begin with she tried to follow,' Fire explained. 'But when she reached the limit of her comfort zone, she panicked and rolled her wings level. Because she had lost so much speed, she had to accelerate to stay airborne, but this powered her ahead of us. As I completed the roll, she was below and in front of me. I could not have asked for a more perfect outcome. Unfortunately, I misjudged the second half of the roll, leaving us in a steeper dive than I anticipated. We nearly flew straight into the ground. The next time we do it, I will be sure to start with more height.'

'NEXT TIME!' Nolita choked, still struggling to believe that she had actually been upside down on her dragon's back. 'There will be no "next time"! No "we"! and no more crazy stunts! Is that clear – NEVER!' She shuddered. 'Shouldn't we be trying to catch up with Pell and Shadow now?' she asked, trying to change the subject.

There was a long pause before Fire answered.

'We can't follow them,' he replied wearily.

'Why not?'

'They have crossed the line of mountains to our right. Unless I can find a low-level way through, or they re-cross to this valley, we're on our own for a while longer. I'm far too tired to consider trying one of the high passes. I will try to put some distance between Deepshade and us, but I must land soon. I am exhausted. I need to rest.'

All the wild emotional swings of the past few minutes faded as the implications of Fire's words sank in. The cold hard core of fear solidified in Nolita's stomach and she began to shake. She had been alone with Firestorm for a few hours in Isaa, but then they had had a plan to reunite with the rest of the group. This time there was no plan. She was alone with her dragon, and that scared her even more than flying upside down had done.

Chapter Eleven
Hatching

'We've already lost a whole week!' Kira snapped, her eyes narrowing. Her whole body was taut with emotion. 'We can't stay any longer, Elian. Make sure you're ready to go. Fang tells me Aurora thinks you're fit enough to fly. We should leave tomorrow at dawn. Fang and I have both been experiencing a growing feeling that the answers we need are in the other world. I know you must be reluctant to go back after our last visit, but we have no choice.'

'Blaze's eggs are likely to begin hatching any day now,' Elian replied. 'Aren't you curious to see hatchling dragons?'

Kira gave him a hard stare. She could see it was a shame to miss such a unique occasion, but she was determined. 'Yes, I'm curious,' she admitted. 'But I'm not going to start cooing over a gaggle of baby dragons and allow them to sidetrack me from our quest. We're running out of time and we're no closer to finding the final orb now than we were when we arrived.'

There had been a growing sense of anticipation as the eggs had hardened during the last few days. Tarl and Neema had fussed about the enclave, ensuring there was plenty of food ready for the hatchlings and preparing the caves where the newly hatched dragons could spend time until they were ready to fly out and fend for themselves.

If the dragons could have used their power to make Elian better, they would have left sooner and his fascination with the hatching would not have been so acute. Shimmer had been terribly disappointed when his attempts at healing Elian had failed. He did exactly as Aurora had directed, but nothing happened.

Aurora had also tried to heal her rider – to no avail. Unlike a day dragon's potent healing fire, it appeared that while a dawn dragon's ability to transfer its inner power through its scales could be used to heal dragons, it could not be used to heal their human riders.

Elian's eyesight had continued to recover naturally during the week. It was almost back to normal now, with only a slight blurriness, but his hearing remained limited. The background ringing noise had gone, but he was still almost completely deaf in his right ear.

The wound in Elian's leg was also still painful. The damaged muscles in his right thigh were not yet able to bear his full weight. Where Haithan had removed the fragment of metal, the flesh appeared to be healing cleanly, but it required bandaging with linen strips twice a day, and would continue to do so for some time.

Elian had progressed quickly from hobbling around the enclave on wooden crutches, to limping with a single stick for support but, unless they could find a friendly day dragon to cut short the healing process, the leg was going to take a long time to fully recover. He was reluctant to get going, but he knew in his heart that he was fit enough to fly and they could not delay any longer.

Finally, Elian sighed and nodded. 'Do you know where to look in the other world?' he asked Kira. 'I don't want to stumble around blindly. To be frank, I don't want to go at all.'

An intense flashback to the explosion caused his chest to tighten. For a moment he was there again. It was dark. The strange, remote battle was going on around him. He was blind, deaf and paralysed by fear. Then it was gone. His eyes refocused on Kira and he gave an awkward grimace as he fought down the remnants of panic that threatened to overwhelm him. He took a slow, deep breath. Then another.

'I don't think we'll have to search,' she replied carefully. 'I've got a feeling the answer we're looking for will find us. The man, Jack Miller, is the key. I don't pretend to understand what his connection is, or why he is a part of this, but we need to find him again if we want to find the orb. Fang and I can both feel it. Aurora feels it too. Talk to her. She's been hesitant to push you into going back because she shares your apprehension, but she feels the pull as we do.'

Elian did not need to talk with Aurora to know that Kira was right. He could feel it too, but the thought of returning to the other world terrified him. The adventure of riding a dragon and embarking on a quest had been exciting at the beginning. Faced with the prospect of going back to the dangers and horrors of a war he did not understand sent wave after wave of icy fear down his back.

'We're not furthering the quest by staying here,' she added, watching him intently. 'The hunters in my tribe would not have stayed in one place this long unless they were convinced of eventual success.'

Elian felt his face begin to burn with embarrassment. Did she think him a coward? At the moment he felt like one. Nolita had faced her fears. Was this how she had felt before her trial in the Chamber of the Sun's Steps? If so, then he now appreciated just how brave she was.

'*You are not a coward, Elian,*' Aurora assured him. '*You are nervous. That is understandable. I am nervous, too.*'

'Go get some sleep, Kira,' he advised, turning to busy himself with packing things into his saddle-bags in an effort to find focus and hide his flushed cheeks. 'Don't worry. I'll be ready. I'll be turning in shortly. Can you ask Neema and Tarl to come and see me, please? I'd like to thank them before we go, but I don't want to wake them before sunrise to do it.'

'Of course.'

Kira strode away quickly. The fear in Elian's eyes had been obvious. She did not want to embarrass him unnecessarily. He had suffered a trauma that would take time to heal, but they did not have the luxury of giving him time to recover fully. They had to complete the quest with all possible speed. Segun had had another week to marshal his dragons. How many were blockading the Oracle's cave now? And what about Nolita and Pell? Had they reached the day dragon enclave yet? Would Barnabas bring the day dragons to their aide? She hoped so. If not, the job of trying to sneak past Segun's men would surely fall to her and Fang. While normally she was happy to brave danger, she felt that a solo effort would be doomed to failure. Segun was no fool. With the resources he had available, she doubted he would leave any chink in his defences for her to exploit.

Tarl and Neema were hovering at the entrance to the hatching chamber. Kira could hear Blaze shuffling around inside, the dragon's great talons clicking and scraping against the stone floor. They agreed to go and see Elian straight away, but Tarl gave an anxious glance back towards the chamber as he left, and Kira had to work hard to keep from smiling at his worry. He looked shattered through lack of sleep and she doubted that tonight would bring him the rest he needed.

Her message delivered, Kira went to her chamber, packed her saddlebags ready for the morning and climbed into bed. Excited by the prospect of getting underway again, Kira found it hard to get to sleep at first. When she finally drifted off, it seemed she had hardly settled when Fang woke her. She groaned. Surely dawn could not be here already?

'No,' he said. *'It's not time to leave yet. We still have several hours until dawn, but I thought you would want to see this before we go.'*

'*The eggs?*' Kira asked, jolting upright on the low bed.

'*The first is hatching now,*' Fang confirmed.

For all her restlessness and desire to press on with the quest, the excitement in Fang's tone was infectious. The unique nature of this opportunity was not lost on her, despite her hard line with Elian. She paused only long enough to throw on some clothing before racing out of her room and through the enclave to the hatching chamber entrance. She arrived, breathless, but fully awake a few moments later.

Tarl and Neema were whispering excitedly to one another. They beamed at Kira.

'Didn't think it would be long before we saw you,' Neema said in a breathy voice. 'You're just in time, but don't enter the chamber. When a hatchling is born, all it thinks about is eating. I wouldn't want a dragonet to see you as its first meal.'

'Have any of them hatched yet?'

'The first has cracked its shell,' Tarl said proudly. 'It should break out any moment now.'

'And because it has hatched before sunrise, it will be a night dragon ...' Kira observed, unable to stop a shiver running up and down her spine as she said it.

'Eventually, yes,' Neema replied. 'But according to

131

the archives, when dragons first hatch, they all appear much the same. It takes some time before their bodies change and develop the characteristic nature of their mature dragon type.'

'According to the archives' was a phrase that was beginning to get on Kira's nerves. Everything these two dragonriders did seemed to be referenced to what the archives told them.

'I didn't realise that,' she admitted. 'So are you expecting them all to hatch at about the same time, or will it take a while?'

Neema and Tarl looked at one another questioningly and then shrugged. 'We don't know,' Tarl answered. 'The records don't say. We think it will take at least a few hours for the full clutch to hatch. Take a look. The one that's cracking is towards the right of the pile.'

The three of them arranged themselves in the doorway so they could all look into the chamber without crossing the threshold. The only source of light inside the chamber was Blaze, whose glowing scales bathed the huge mottled eggs in a warm, golden haze. Blaze had curled protectively in a semi-circle behind the eggs. The dragon's eyes were gleaming as she watched intently over them.

A slight movement and a surprisingly loud cracking sound drew Kira's attention. One of the eggs towards the right of the clutch was rocking and a long crack had appeared in its side. Another crack. Another. And suddenly, the shell split wide, spilling an ugly, gangling creature abruptly onto the floor. It screeched with protest and flopped around awkwardly, trying to get to its feet.

Kira felt an almost overpowering urge to run to the dragonet's aide. It looked so helpless.

'Don't!' Fang warned in her mind. 'It will kill you if you get too close. Do not fear. It will right itself soon enough.'

No sooner had Fang spoken in her mind than the baby dragon found its feet. Its oversized head swayed on its long, spindly neck and its legs wobbled as it fought to stay upright. The strangely colourless eyes looked devoid of intelligence. They were cold and flat, and held a look that Kira recognised instantly – the look of a predator.

The dragonet's scales were also colourless. They had an almost translucent quality, though Kira could not actually see the tissue and organs beneath. If she had been asked to describe its colour, she would have instinctively said grey, though it was nothing like the grey of her dusk dragon.

'Why doesn't Blaze help it?' she asked Fang.

'Dragonets must learn to fend for themselves from the moment they are born,' he replied. 'Blaze has put food in

the chamber. Finding it will be the young dragon's first
quest. Do not fret. It will not take long.'

Almost as if the dragonet had heard Fang's words,
it took its first few tottering steps. Arching its neck,
the creature raised its wedge-shaped head high
and took a long, wobbling look around the chamber.
Its gaze swept past the pile of raw meat stacked
not far away and when it moved again, it moved
unswervingly towards the ready food.

'Did I miss much?'

Kira started. She had not heard Elian approach.

'The first one's just hatched,' she told him
excitedly. 'Here. Take a look.'

She moved aside to give him room to lean against
the doorframe. He looked into the chamber just in
time to see the top of a second egg erupt as the
dragonet inside forced its head through the top. It
looked absurd with a cap of shell on its head, but
Kira resisted the urge to giggle. She did not want to
upset Tarl.

The first dragonet was attacking the pile of meat
with startling ferocity. Picking up piece after piece,
it tipped back its head and swallowed each one
whole. It had problems with one of the larger pieces,
coughing it back up several times. But rather than
drop it and move on to another piece, the young

dragon persevered until it succeeded in swallowing the huge hunk of flesh and keeping it down.

'Obstinate little nipper, isn't he?' Kira observed aloud.

'Remind you of anyone?' Elian said.

'You, actually!'

'Ha! Very funny.'

Kira knew that dragons did not really chew their food. Fang sometimes held a piece of meat in his mouth and rolled it around on his tongue, giving the impression he was chewing, but this was an illusion. Dragons ripped a kill to pieces with their sharp teeth and then swallowed the pieces, bones and all. The pile of meat near the eggs had already been shredded, presumably by Blaze, into much smaller pieces than an adult dragon would consume.

The second hatchling shouldered out of its egg and tottered forwards on uncertain feet. Its head turned, eyes scanning the room until it was looking at the doorway, where the four riders were standing. Opening its mouth wide, it let out a screech and took two rapid steps forwards.

Kira gasped as she saw the intent in its eyes. She knew the look from predators she had encountered in the savannah. It was the look of a killer. Although newly hatched, the dragon had nearly as much body mass as she did. It was slow and uncoordinated but, nevertheless, she could appreciate the potential danger it posed.

A rumbling sound from Blaze stopped the hatchling in its tracks. Its head snapped around and it let out a defiant screech to its mother, but the objection was over-ruled by another, slightly louder warning rumble. The dragonet screeched again, but turned obediently to where the first was feasting and staggered the short distance to the waiting food.

Kira realised she had been holding her breath. She let it out in a long silent sigh of relief.

'I think that was our cue to leave you to it,' she announced to Neema and Tarl. 'Congratulations. We'll stop by again before we leave, but we have a potentially difficult day ahead.'

'Of course,' Neema said quickly. 'Don't let us keep you from your rest. It's been a pleasure to meet both of you. Make sure you stop by the enclave once your quest is complete. Tarl and I will be here for the rest of the year.'

'Come and see us before you leave,' Tarl added. 'And we'll say goodbye properly then.'

Chapter Twelve

Fright and Fight

'There! On the left!' Pell exclaimed aloud. 'How far does it go? They're closing fast!'

'There's only one way to find out,' Shadow replied.

Banking hard, Shadow turned into the narrow canyon and raced between the towering walls of rock. Pell wanted to look back to see if Longtail had followed them in, but he could not look anywhere except ahead. The slot they had flown into barely qualified as a canyon. It was about three dragon wingspans wide at the opening, narrowing quickly to two.

A surging stream ran beneath them, foaming across the rocks in a mackerel-striped race of dark water and frothing white. The gushing sound of the mountain stream and the rushing wind added to the sensation of speed as the rock walls flashed past on either side.

'Are they still behind us?'

'Yes,' Shadow affirmed. 'But they hesitated before committing. It cost them their height advantage.'

'Any idea where this canyon leads?'

'I don't think it goes anywhere,' Shadow replied. 'My senses tell me it's a box canyon.'

'Better and better!' Pell muttered. 'Are you ready?'

'Ready as I'm going to be,' Shadow confirmed, tipping first right and then left as she followed the weaving path of the canyon. 'Hold on tight. Here we go.'

As Shadow rolled her wings level, she suddenly pitched hard upwards in an attempted repeat of the manoeuvre she had performed in the mountains of Isaa. The last time they had done this it had saved them from certain death in The Knife – a similarly narrow canyon that they had been tricked into entering.

Up, up, up, Shadow pitched until she was flying vertically upwards. She intended to roll ninety degrees to the left and use her talons against the cliff wall to turn her body until it was facing vertically downwards again. Whether she did not have the momentum this time, or she did not pitch up sharply enough into the vertical was not clear. Whichever it was, the result was not what she planned.

Shadow ran out of momentum before she

completed the quarter turn, or got close enough to the cliff wall to use it to aid turning her body. Pell's memory had softened the fear he had experienced when reaching the peak of their climb in The Knife. The sensation this time was nothing short of terrifying. His bottom floated away from the saddle and his grip suddenly felt weak and uncertain. Worse, he could feel Shadow beginning to panic.

The manoeuvre had gone horribly wrong. They peaked and Shadow began to fall out of control – initially tail first, though this was only momentary. The shape of the dragon's body was not compatible with travelling through the air backwards, and with breathtaking suddenness the airflow flipped her over. The force created by the rapid rotation slammed Pell forwards. His head connected hard with the ridge of dragon horn in front of the saddle, the impact setting the world alight in a swimming golden haze. His grip on the pommel failed and both dragon and rider plummeted headfirst towards the foaming stream below.

Unfurling just one wing, Shadow gave a single flap and spun her body to face back the way she had come. The rotation tipped Pell to the left and, as Shadow abruptly began to pitch out of the death dive, he slid down her side until he was hanging by his feet from the stirrups.

'PELL!'

Shadow's shout in his mind snapped him out of his semi-conscious state. The next few heartbeats were agony. Shadow pitched up hard to recover from the dive. The force dragging Pell downwards became immense. Only his stirrups saved him from certain death. Had both of his feet not been firmly wedged in the metal D-shaped footrests he would have fallen. Also, had he been sitting upright in the saddle as Shadow bottomed out of the dive, the night dragon coming in the other direction would have struck him.

What Shadow's diving run must have looked like to Longtail and her rider, Pell could only wonder. They must have thought Shadow had made the spectacular reversal in order to turn from defence to attack. For the briefest heartbeat, the two dragons flew head to head and looked like colliding with a high-speed closure that would certainly have proved fatal to all. At the very last moment, Longtail climbed just enough to pass over Shadow.

Even though the blow to his head had dazed him, Pell, hanging upside down, was perfectly placed to appreciate how close the dragons came to hitting one another. For the briefest moment, the night dragon's wings overhead eclipsed all else and Pell caught a glimpse of her great talons as they passed through

140

the place he should have been sitting. There was a whoosh of air that dragged at him for an instant and then it was gone.

His head was thumping with a combination of the impact and the extra blood that was pounding in his ears from being suspended upside down.

'I'm still with you, Shadow,' he groaned.

'*Can you get back in the saddle?*' she asked, her voice sounding both relieved and worried. '*Tell me if you're likely to fall and I'll land.*'

'*I'll try,*' he promised, the words reverberating through his head.

The rushing water and jagged rocks were whizzing past. He was not sure if the sensation of extreme speed was real, or an illusion created through being close to the ground and inverted. Either way, he had no desire to fall. Tightening his stomach muscles and pushing his bottom against Shadow's side, he folded upwards and reached for the stirrup straps. He caught the left one first time and, with a strong grasp, he wriggled and pulled his way back up into the saddle.

The exit into the main valley was visible ahead. There was no sign of Longtail. Despite pain that made his skull feel ready to explode, Pell grinned. They had done it. By the time Longtail worked out a safe way of turning around, he and Shadow would be gone.

Even if Longtail had landed and turned around immediately, there was no way she could take off again in there. The stream dominated the base of the gorge and a dragon needed a good run on firm footing to take off. She would have to walk out into the main valley to get airborne again, which gave Pell and Shadow the time they needed to escape.

'Are you all right, Pell?'

'I've got a splitting headache and an impressive lump coming up on my forehead,' he said through the bond, gently running a finger over the rapidly swelling area. 'But aside from that I'm fine.'

'I don't think I'll be attempting that again in a hurry,' Shadow confessed. 'I thought I'd lost you when I ran out of energy and flipped over.'

'That part was . . . unexpected,' Pell said. 'But the end result was what we intended.'

'I think I'd rather fight against the odds than repeat the sensation of falling. I've never felt out of control in the air before. Even as a dragonet, I was always a strong flyer.'

Pell did not respond. He could feel how deeply the experience in the gorge had affected her. If he had been asked to give a single defining characteristic of his dragon's mind, he would have said 'confidence', but feeling her sudden uncertainty made her feel like a stranger. It reminded him of the first time they had met and the shock he had felt when he looked into her mind through the bond. As well as the pervading weariness, her personality felt somehow diminished.

They emerged from the narrow canyon into the expanse of the main valley. Pell saw Darkhorn almost instantly. She was approaching from the right, flying along the middle of the valley and apparently unaware of their presence.

'*I see them,*' Shadow told him, her voice carrying the strange echoing quality it had when she was shielding her thoughts. '*Hold on tight. I am done with running. It's time we went on the offensive.*'

Shadow began climbing again and Pell felt his stomach tighten with anticipation. Darkhorn held a slight height advantage, but Shadow eliminated it within a few wingbeats. They were closing on a rapid intercept course and already well above Darkhorn when she saw them. The night dragon lurched into a turn towards them, but then seemed to have second thoughts. She twitched first one way and then the other, no doubt wondering what had become of Longtail.

Unswerving, Shadow arrowed forwards in a head-on charge. It suddenly occurred to Pell that he had no idea how dragons fought in the air. What would Shadow do? He could feel her muscles bunch beneath him as she braced herself for battle. Darkhorn bloomed ahead of them and Shadow reared to strike even as her opponent did the same. Both dragons screeched as they collided, becoming a frenzy of slashing talons and teeth.

The fight was vicious and short. To Pell it felt almost like a dance. The two dragons turned hard in a tight spiral, wings beating hard to limit their rate of descent, but both spinning downwards in a slashing, biting wrestling match. Shadow gained the advantage quickly. She was bigger than Darkhorn, and used her size and extra weight to force her opponent down.

Both dragons were fighting to maintain their height above the ground. Both were losing. They had not been high when the fight began, but what altitude they had enjoyed disappeared fast. Until the fight began, Pell could not take his eyes off Darkhorn. Once they had engaged, he found it hard to tear his gaze from the expanse of trees in the valley as it swelled rapidly towards them.

With a roar of triumph, Shadow gained a hold on Darkhorn's neck. First with her teeth and then with her front talons, she took control. Twisting to arch over the smaller dragon she gave a mighty shove downwards, gaining the boost she needed to begin climbing again. The sound of cracking, splintering wood told part of the story, but Pell was not in a position to see the results of Shadow's victory.

Almost brushing the treetops, Shadow laboured

to stay in the air. Pell's heart was drumming double-time, as he willed her to find the strength she needed to stay clear of the grasping trees. Slowly, span-by-span, she dragged herself back into the sky, accelerating to a more normal flying speed as she went. Once they were well clear of the treetops she made a gentle turn that allowed Pell to look back. There was no sign of Darkhorn. It was as if the trees had grabbed her and swallowed her whole.

'She is alive and her rider is unharmed,' Shadow informed him. 'Though I don't think she'll be in a hurry to fly again.'

'Good!' Pell responded. 'Now let's get away from here before Longtail finds her way out of that canyon.'

Shadow did not reply, but he sensed that she felt the fight had already delayed them too long for a clean escape. They completed their turn and rolled out heading along the valley that took them deeper into the mountains. Pell watched the entrance to the narrow canyon anxiously for signs of Longtail. Sure enough she emerged before they had got out of a direct line of sight.

'They have seen us,' Shadow confirmed. 'But it appears they have decided to give up the chase. They are talking with their friends in the trees.'

That was a conversation Pell wished he could hear, but he was content to have defeated them. The lump on his forehead was pulsating with pain and he felt sure he had pulled several muscles in his arms and back. All he wanted to do was find somewhere to rest where they would not be found, but his conscience would not let him rest yet. His promise to Elian haunted him. Somehow they had to find Nolita and Firestorm.

The valley curved gently left and it was not long before the night dragons were far from view.

'We'd better start looking for a way to loop back towards the others without having to cross the ridge,' Pell suggested.

'Very well,' Shadow replied.

Their thoughts on what they would find went unspoken, but Pell could tell that his dragon's thoughts echoed his own. She did not expect to find them alive, which meant they were likely to face more trouble from Deepshade.

Chapter Thirteen
Finding Jack

Getting into the saddle was not easy, but with Kira's help Elian overcame the difficulties of climbing up Aurora's side. Back in the saddle, he felt an overwhelming sense of relief. It was as if he had been lost for a long time and finally found his way home. He shuffled his bottom around until the pressure on his right thigh was minimal.

He looked down at Kira. 'Thanks,' he said. 'I'm all set. You'd better mount up. I can feel the dawn approaching.'

She nodded and ran across to Fang, who was waiting patiently nearby. With nimble agility she leaped up into her saddle.

Elian and Kira had said their goodbyes to Neema and Tarl. A dozen dragonets had already broken free from their shells and more eggs looked ready to crack at any time. It was an exciting time and busy for the two dawn dragonriders.

'Are you all right, Elian?' Aurora asked. 'I can feel your discomfort.'

'I'll be fine,' he replied. 'It's going to be painful, but I'll cope.'

Elian could not voice the fears that plagued him. What horrors awaited them on the other side of the gateway this time? Their last visit had left him with injuries that he doubted would ever heal. An icy feeling in the pit of his stomach made him feel sick with worry.

Aurora gave the mental equivalent of a sigh. 'I do,' she admitted. 'Visions of a world without the Oracle plague my thoughts. We must find the final orb soon, or all our efforts will have been in vain. Do not fear, Elian. I will protect you to the best of my ability. Whatever awaits us, we shall face together.'

'Then let's go and get this over with,' he urged. 'Let's find Jack and see if he's got any ideas.'

Aurora moved gently until she was in position to run towards the mouth of the cavern. Outside the cave the slope was steep. A strip down the mountainside had been cleared of all large rocks. This gave the dragons some distance to accelerate to a safe flying speed before launching.

Elian winced as he settled into a balanced position

148

for the take off. The pain in his leg was intense, but he gritted his teeth and did his best to ignore it. Aurora did not need to give him further warning before leaping forward towards the mouth of the cave. With her mind linked to his, she sensed him reading the play of her muscles as she prepared and knew he was ready for the explosive acceleration. She raced towards the opening, her talons clacking with increasing tempo on the stone floor as she picked up speed. The instant they emerged into the open she spread her wings and with one final push of her hind legs she launched into the air.

As they left the ground, Elian experienced a momentary dizziness that rocked him in the saddle, leaving him feeling nauseous and struggling to stay upright. It was not quite the same as the twisting disorientation he had experienced when entering and leaving Aurora's gateways between worlds, but the similarity was disturbing. He took several deep breaths, drawing them in and expelling them slowly. Gradually the feeling subsided until he was able to enjoy the sensation of flying again.

A light drizzle hung in the air, the tiny droplets of water seeming to float towards the ground rather than fall. The atmosphere felt still and heavy. It was murky with a dull grey veil of cloud shrouding the mountaintops around them.

Elian had hoped to see something of the area around the enclave, but all he could see was that they were in mountainous terrain. It was warmer than it had been in Isaa and the smells that pervaded were different from those of Orupee, but the scenery was similar to all mountainous areas he had seen during the past month. There was no time to explore. He could feel the moment of dawn approaching.

Aurora began to draw on the energy the rising sun gave her. Through the bond he could feel she was strong and rested. Power flowed easily and he felt the gateway opening ahead of them. The grey of the swirling vortex was barely visible against the murky background, but Elian could feel exactly where it was. They slowed, allowing Fang and Kira to overtake them and lead the way in. The gateway swallowed the dusk dragon and his rider even as Elian felt a surge of joy rush through Aurora's mind. Before he could ask her about it they plunged into the void.

The twisting sensation of entry did not affect Elian as much as the disorientation he had felt during the take-off run. Floating in the grey nothingness between worlds brought no fear, but as the wrench of emergence signalled their arrival he felt his stomach tighten.

It was dark again. Memories of his most recent

night experience over the battle lines assaulted him. A lance of pain shot through his injured thigh as his tightening muscles put pressure on the wound. He hunched over the pommel, shifting his position in the saddle to try to relieve the pressure on his right leg.

'*Are you all right, Elian?*' Aurora asked, feeling his discomfort.

'*To be truthful, no, Ra,*' he admitted. '*But I'll survive. Where are we this time? Can you tell?*'

Distant flashes to their right had already given him the clue that they were not over the battle lines, for which he was most grateful.

'*We seem to have emerged near the small area of woodland where we last saw Jack Miller,*' she said. '*Should we land there, do you think?*'

Elian thought for a moment. '*Jack appeared keen for us to avoid the trap in the other woods where we were attacked by the dogs,*' he replied. '*So, I imagine he will have tried to avoid telling his superiors about this place. Can you feel him anywhere nearby?*'

The sensation as Aurora reached out with her mind felt more intense than he remembered. It was almost as if he was reaching out with her, scouring the sky for a familiar sense of presence. He knew the answer before she gave it.

'*No,*' she said. '*At least, he's not in the air. He could

be on the ground nearby. The field where he flew his machine from is not far from the trees. Perhaps we should land here, use the trees as cover, and begin our search by looking there.'

'That makes sense.'

Aurora entered a gentle spiral descent. Elian could see no sign of Fang and Kira. The dusk dragon would most likely have camouflaged the moment he emerged. Given his natural dark-grey colour, he would be difficult enough to pick out at night under normal circumstances. With his camouflage employed, he became totally invisible.

A broken layer of high cloud obscured the large silver moon, but the glow of it could still be seen. It took a little while for Elian's eyes to adjust, but gradually he began to distinguish between the different shades of darkness below. He picked out the stand of woods some time before they began their approach to land. Aurora swept the area with her mind, but felt no sign of human life amongst the trees, or in the immediate vicinity of their chosen landing spot.

When they touched down, Elian remained mounted until Aurora had eased between the trees. No sooner had he begun to try climbing down than Kira was there to help. As was usual with Kira, he had not heard her coming and jumped at her touch.

He could see her grin at his response, but he was more grateful for her help than annoyed by her talent for silent movement.

Aurora led them to the place where he and Pell had awaited the dawn with Jack the last time they had seen him.

'I think I can feel Jack now,' Aurora announced suddenly. 'But he's not alone. There is a group of buildings not far from the eastern side of the wood. You will need to cross an open field, but you should be able to get there unseen. I can direct you, if you wish. I was going to suggest you wait until early morning, but the hour is not as late as I first thought. My only concern is the pain in your leg. Perhaps Kira should go.'

'No, I'll do it,' Elian insisted, determined to put aside his fears. 'There's no telling how the fighting men of this world will welcome a dark-skinned girl who paints her face. Kira is brilliant at stalking, but I think it'll be better if I do this. I promise I'll go slowly.'

Kira was not happy at being left behind but, to his surprise, she listened to his logic and offered little argument. Her eyes were flashing with contained anger and concern as he hobbled away into the darkness, although she did agree not to follow him unless Aurora relayed through Fang that he was in trouble. As he left, she began gathering materials for a shelter.

Elian was far from silent moving between the trees, but once he reached the far edge of the woods he could hear the distant sound of music and singing. It appeared to be coming from the direction Aurora had told him to take. No one was likely to hear him approach over the sound of the music, so he limped as quickly as he could across the open field to the hedge on the other side.

He could hear people talking and laughing now. They were not far away. He slid along the line of a hedge towards them. As he drew nearer he stayed low, remaining deep in the shadows. Nervous at how he would be received, Elian began thinking about what he would give for Kira's skill at moving silently. He would be much happier if he could get to Jack without being seen by anyone else.

'Don't worry, Elian. You're doing fine, but stay where you are for a moment and get down low. There's something approaching you at speed.' Aurora's warning gave Elian the distinct impression that she was curious.

'What do you mean something?' he replied. A feeling of apprehension set the hairs on the back of his neck prickling and a rash of goosebumps rose on his arms.

Aurora did not answer, but a distant rumbling growl sent a shiver of fear through him. Whatever the thing was, it sounded big. What was more, it was moving fast – really fast! The distant rumble transformed into a throaty roar as the strange beast closed rapidly on his hiding place.

In a moment of rash bravery, Elian got to his feet and peered over the bushes to see if he could catch a glimpse of its advance. There was a narrow lane on the other side of the hedgerow. To Elian's amazement he could see something huge with two blazing eyes, whose gaze cast beams of bright, white light wherever they looked. It was racing towards him at the speed of a galloping horse. Fear, together with the dark's exaggerating properties, swelled the creature to impossible looming proportions. For a moment he froze as the great eyes swung their bright focus in an arc towards him.

'Elian!'

The warning was timely. He ducked and placed his arms protectively over his head as the noise of its approach swelled to fill his ears. Strange rattling and squeaking noises were apparent above the throaty roar as the beast sped by no more than a few paces from where Elian cowered. It was the metallic squeak that made him stand up and look around. Jumping to his feet, he looked over the hedgerow again just in time to make out the shape of the behemoth.

A waft of oily fumes confirmed Elian's conclusion. '*It's a horseless wagon! First flying machines and* now this! The people here are certainly clever with their inventions.'

'Indeed, Elian,' Aurora agreed. 'Perhaps too clever for their own good. What have their ingenious machines gained them? I would not be surprised to learn that the fighting is linked to a drive for invention. The people of this world have had many conflicts, but this is unlike any I've seen.'

'I'm not sure I want to know more about the fighting,' Elian said. 'I just want to find out where the final orb is and get out of here.'

The horseless wagon had squealed to a noisy stop just a few hundred paces further up the lane. Elian pushed through the hedge and ran silently towards it. He could see men emerging from a door in its side and walking up a short path to a nearby building. No one looked in his direction. Their attention was on the building ahead and talking.

'Is Jack in there?' Elian asked apprehensively.

'No. There's another building about a hundred paces behind that one. He's there. Try to get through the hedge on the other side of the lane. The field beyond will take you closer.'

'You're beginning to sound faint, Ra,' Elian noted anxiously.

'Don't fret, Elian. I can still hear you and there's not far to go.'

156

This hedgerow was thicker and more difficult to traverse than the previous one. Brambles clung and tore stubbornly at his clothing as he forced his way through. He seemed to be making an inordinate amount of noise, but he was counting on the now chugging breath of the stationary horseless wagon to blanket any sound he made. He was through and on the move again. He could see the house. Aurora was right – it was not far. There was a gate leading from the field to a courtyard by the house. He raced across the open ground to it, trying his best to keep his footfalls light.

Crouching by the gatepost, he scanned the courtyard. No one moved. Music and laughter were emanating from within the house. No light escaped from the windows, but whatever was blocking the light could not totally dampen the sounds of a party. How could people sing and laugh with the grumble of battle in the background?

Slipping between the bars of the gate, he hobbled as fast as he could across the courtyard to the side of the building. The windows had been covered inside with something more than just material drapes. He peered through the glass, hoping to catch a glimpse of what was going on inside, but whatever had been used to black out the windows was too effective.

Frustrated, Elian stepped into the porch and

listened at the door. Aside from the music, there was a buzz of conversation. It was hard to pick out individual voices. He pressed closer to the door, hoping to hear a little better. Taut with concentration, he jumped as a large hand suddenly clamped onto his shoulder with an iron grip.

'And what do you think you're doing, lad?'

Chapter Fourteen
The Proposition

The soldier's voice was gruff and angry. In the dim moonlight, his silhouetted figure looked huge.

'N . . . n . . . nothing, sir,' Elian stammered, his tongue tangling over his words as he tried to combat the shock. 'Well . . . that is, I'd like to speak to Jack Miller, if it's possible.'

'*Jack* Miller, is it? A friend is he, lad? Expecting yer perhaps?'

'Not a friend exactly, but we've met before. Can you ask him if he'll speak to me? He's not expecting me tonight, but I'm sure he'll want to see me. Tell him Elian is here.'

It was hard to see anything of the soldier's expression. His face was hidden by the shadow of his steel helmet. The man was still for a moment, as if trapped by indecision.

'Very well, lad, I'll see if Captain Miller is willing to talk with yer. If not, then yer going to spend an uncomfortable night in the cellar, followed by a visit to the old man tomorrow. Yer can explain to him what yer business is, sneaking around. There's been plenty shot for spying, so yer'd better make sure yer excuse is a good 'un.'

Elian was not sure whether the man was being serious, but his deep voice sounded convincing in the dark porch of the farmhouse.

'Why didn't you warn me he was coming?' Elian asked Aurora silently, mentally cursing his impaired hearing for not picking up the man's approach.

'Sorry, I didn't sense his presence until it was too late,' she answered. *'Don't worry. If Jack does not come, I'll get Fang to send Kira. She'll find a way to help you escape.'*

'Great,' Elian said, trying not to lace the word with too much sarcasm. It was not that he did not trust Kira to get him out – the opposite was true. What irked him was that he might be reliant on her again to rescue him from captivity.

The soldier took a firm hold of Elian's shirt at the back of the neck and rapped on the door with the thick end of the weapon he was holding in his other hand. Elian made no attempt to struggle. His eyes followed the weapon. He had taken it for a club at

160

first, but it was not shaped like any he had ever seen before. He could not see it clearly in the dark, but he could tell it was not pointed enough to be used like a spear. Neither was it balanced like a staff, or shaped to swing like a club.

The door opened suddenly, golden light and a waft of warm air spilling out onto the threshold.

'Yes? Is there a problem, Private?'

The first thing that struck Elian about the man in the doorway was his clothing. He had never seen such garments before. The shiny line of buttons, the four large rectangular pockets on the close-fitting jacket and a curious winged symbol above the top left pocket looked very smart.

'The boy here wants to see Captain Miller, sir.'

Elian looked the man in the eyes as he was appraised. He only looked to have seen about eighteen season rotations, yet the older soldier gripping Elian's collar had called him 'sir'. Was he some sort of nobleman?

'What do you want with Jack, sonny?'

There was amusement and curiosity in the man's eyes as he waited for Elian to answer. Elian was not sure what to tell him. He did not think telling him he was a dragonrider would be a good idea, but he needed to be sure the man would take him seriously.

'I need to talk with him privately, sir,' he replied

cautiously. 'The last time I saw Jack he insisted the subject of our discussions remain secret. My name is Elian. Can you tell him I'm here, please? My quest is urgent.'

'Quest, is it?' the man said, thoughtfully stroking his neatly trimmed moustache with his right index finger. 'You have a strange accent, lad. Where are you from?'

'I'm afraid I can't tell you that either, sir.'

'Is that so? Very well. Keep him here, Private. I'll go and see what Captain Miller has to say on the matter. I shouldn't be more than a minute or two.'

The door closed, plunging the porch into darkness. After looking into such a bright source of light, Elian found his night vision had gone and he was all but blind. The soldier's grip on his collar loosened slightly, but he did not let go.

'Who was that?' Elian asked. 'Was he some sort of lord?'

'A lord?' the man grunted, giving a gruff bark of laughter. 'Acts like one, fer sure. Most of the officers do, lad. I'm surprised yer don't know that. Miller's no different from the rest. If yer making this story up about knowing him, yer about to find yerself in a whole mess of trouble.'

Elian briefly considered responding, but thought better of it. The soldier was not going to believe him, no matter what he said. He barely had time to think this before the door burst open again and there was Jack. He was dressed in almost identical clothes to the officer who had opened the door the first time. His face was unmistakeably excited.

'Elian!' he exclaimed. 'That's wizard! Thank you, Private. You may let him go. I'll look after him now.'

'Very good, sir,' the soldier replied. His tone was businesslike, though Elian was sure he detected an undertone of disgruntlement. The hand let go of his collar and the man suddenly stood very straight and raised his right hand to his forehead. Jack repeated the gesture and the soldier, having been dismissed, did an abrupt about turn and left.

'It's been so long that I thought you might not come back,' Jack said in a low voice, drawing Elian in through the door and into the warm hallway. The other officer was watching from a few paces away. Jack suddenly noticed him. 'Pete, can you do me a favour?' he asked.

'That depends, Jack,' he replied. 'What do you want?'

'Can you drag the old man from the piano and tell him to meet me in the back room? I need to speak with him urgently.'

'Seems everything is urgent tonight,' Pete said, raising one eyebrow quizzically. 'He won't be happy,

Jack. You know how he is when he gets singing.'

'Well let that be my problem, Pete.'

'On your head be it then.'

Elian was surprised to see the man Jack called Pete incline his head towards him in a polite gesture of acknowledgement before he turned and entered the room where all the noise was coming from.

'I was injured by one of those flash-bangs that explode in the sky,' Elian told Jack in a low voice. I haven't been able to fly for about a week now.'

'Was it bad?'

'Fragments of metal in my leg, hearing loss and blurred vision,' Elian reported. 'It wasn't nice.'

Jack winced at the catalogue of injuries. 'No,' he said. 'Not pleasant at all. Still, it's good to see you're a survivor. For me, it's been months since we last met. I've been lucky enough not to prang any more kites in that time, but although I'm still in one piece, it's been tough. I've been involved in so many scraps that I'm amazed I'm still around. We lose boys every week. Even the best pilots have been falling to the enemy.'

Elian could see the pain in Jack's eyes. He looked older than Elian remembered. It appeared the constant conflict had affected him deeply.

'Jack, I don't know if you remember what I told you about my quest . . .'

'I do,' he said, his eyes suddenly sparkling. 'The four orbs, the Oracle and the night dragons who are doing their best to stop you – I remember it all.'

'Well, Ra and I think you hold the key to finding the final orb.'

'And you're right,' Jack replied, his eyes twinkling.

Elian's jaw dropped. Had he heard Jack correctly? Did he really know where the final orb was?

'The first orb drew blood,' Jack went on. 'The second was formed from a heart. I take it you've found the third, given that you're looking for the final one. If I solved the Oracle's riddles correctly, then the third orb was an eyeball. Am I right?'

'Yes . . . but how?'

'I said I was good at puzzles,' Jack said with a grin. 'The mystery of the poem was rather obvious when I looked at it the right way.'

'So what is the last orb?' Elian asked, his chest tight with excitement.

'I'll tell you in good time, old boy. Don't fret. But first I need to know where your dragon is and if there are any others with you?'

'She's in the woods where we left you last time,' Elian replied, unsure exactly what Jack meant by 'old boy'. 'Kira and Fang are with her. Please, tell me. I must find it quickly. We're almost out of time.'

'Kira is the girl with the invisible dragon, isn't she?' Jack asked, ignoring Elian's plea.

'That's right.'

'Wizard!' Jack breathed. 'Absolutely wizard!' Suddenly his voice became more urgent. 'What about Pell?' he asked.

'He's gone with Nolita and Firestorm on a journey to the day dragon enclave,' Elian said warily. 'The night dragons have blockaded the entrance to the Oracle's cave. We think the only way we're going to get past them is with a show of strength.'

'Interesting. Ah! Here comes the boss. Come, Elian. We've got a lot to talk about.'

Elian was confused. If Jack knew the answer to the final riddle, why was he being so evasive and what was there to talk about? All Elian wanted was the orb. Jack knew that. Why did he want Elian to talk to this 'boss' person?

Jack took Elian by the arm and led him gently across the hall and through a door. Another man followed them, closing the door behind him. The room was full of curiosities – machines that ticked, paintings depicting fascinating scenes and characters, oddments that might have been ornamental, or maybe had purposes that Elian could not instantly determine. Another time Elian could have spent hours poking around in this room, but for the moment he kept his focus on Jack. He wanted answers.

'*Are you still with me?*' he asked Aurora.

'*Yes, Elian,*' she replied.

'*Do you know what's happening?*'

'*Not much more than you do,*' she said. '*But I sense no harmful intent. I suggest you listen to what he has to say. He believes he has what we need. He knew about the third orb, which gives credit to his claim. Let's see what he wants.*'

Elian met the eyes of the newcomer and recognised instantly that he was not the only one with questions. The man seemed to be about four or five season rotations older than Jack. His hair appeared to have been slicked back over his head with some sort of oil and his brown eyes were quick and alert.

'What's this all about, Jack? Who is this boy and why's he here?'

'Boss, meet Elian,' Jack replied. 'Elian, meet Squadron Leader "Wily" White. This, boss, is the boy who's saved my life on more than one occasion. He flies the golden-coloured dragon I told you about last year.'

'*This* is your fabled dragonrider?' White replied, his eyebrows drawing together in a frown as he studied Elian carefully. Elian could see that he was not what the man had expected. 'And I suppose your dragon's waiting outside, lad,' The Squadron Leader said, his lips quirking up the corners of his mouth into a mocking smile. Should he tell? Elian looked across at Jack, asking the question with his eyes. Jack nodded.

'She's in the stand of woods about four hundred paces west of here, sir,' he replied.

'Of course! In the woods,' he replied, his voice laced with sarcasm. 'Did anyone see you arrive, perchance?'

'A man caught me listening at the door, sir,' Elian said hesitantly, unsure of why the Squadron Leader was behaving so strangely. Jack had clearly told him about the dragons, so what was his problem?

'One of the privates on patrol, boss,' Jack explained. 'He didn't see anything, or he would have said.'

Squadron Leader White shook his head impatiently. 'What I meant, Elian, is did anyone see you land on your *dragon*?' he asked pointedly.

'No, sir,' he replied. 'At least, we don't think so.'

'We?'

'Me and my dragon.'

'Ah, yes! You *and* your dragon. Of course.'

Squadron Leader White looked thoughtful for a moment.

'So, has Jack asked you about his grand plan yet?' White asked Elian.

'No, sir,' Jack interjected. 'Not yet. I thought I'd better wait for you.'

'Ask me what?' Elian asked, his suspicions growing rapidly. 'I told you before that we're not going to be drawn into your war, Jack. The dragons will not fight for you, if that's what you're thinking.'

'I haven't forgotten,' Jack said, his face serious. 'But I've got a proposition that I'd like you to consider. It's simple. I'll tell you what you're looking for if you'll help me with one very special mission.'

'I've already saved your life more than once, Jack,' Elian pointed out. 'You're in my debt. This mission you're talking about is dangerous, isn't it? Why should I have to take risks for you again? Please, just tell me where the final orb is.'

'My life isn't that important,' Jack replied with a shrug. 'If I die there'll be a replacement here within a few days. But I know how much value you place on the orb. You told me last time that your world will change for ever if you don't get it to the Oracle in time. Its worth is far greater than my humble life.

'My dragon won't agree to being used as a weapon,' Elian insisted firmly.

'That's not what I'm asking for,' Jack said quickly. 'In brief, I want you to help me capture a particular enemy pilot who has been causing us a lot of trouble. Doing this would give a huge boost to morale amongst our troops. Help me to get him and I'll tell you what you need to know.'

'I don't know . . .' Elian began.

'No one needs to die,' Jack added quickly. 'It's a simple exchange – the pilot for the information. He should be easy enough to find. He flies a bright-red triplane and his name is Baron Manfred von Richthofen.'

Chapter Fifteen
Alone

Even before Firestorm had come to a complete stop, Nolita had unwound the straps from her wrists, thrown her saddlebags clear and slid down his side. She stumbled as she hit the ground. Her legs felt stiff and weak after the long flight, but she could not stagger to the edge of the nearby pine forest fast enough. Desperate to escape her dragon's presence and heedless of scratches to arms and face, she pushed in through the needle-laden branches.

The trees were densely packed, but she continued forcing a way forwards until she could look back without seeing any hint of the blue dragon. Only then did she sink to her knees on the deep carpet of needles. Curling in on herself, she put her head in her hands and began to sob.

Nolita knew she was being irrational. Firestorm

loved her. He wanted to protect her. Given her intimate knowledge of his mind, she could not understand why her intense fear of him remained.

Her terror made no sense. Nolita had flown on Fire's back for weeks, successfully driving back the shadows of her fear enough for her to function as a dragonrider. The darkness had always been there in her mind. Lurking. But she had contained it with her rituals and the support of her friends. Now her friends had gone, leaving a chink in her armour that the dark fears were quick to exploit.

'*Nothing* has changed,' she wept aloud, willing herself to believe the words. 'Fire is the *same* dragon. *Nothing* has changed.'

But something had changed. She was alone. Nightmares of being alone with a dragon had haunted her since she was a little girl: immense, looming, razor-sharp teeth and wicked horns. Despite her determination to suppress her fears, the memory of those dreams sent a shudder through her body.

Nolita had always known Firestorm would come for her. Her mother had believed her nightmares to be a symptom of her fearful nature, but Nolita had always been certain it would happen. She had tried to run and failed. There was nowhere she could hide. Today felt like her trial of bravery in the Chamber of the Sun's Steps all over again, but this time there was no one cheering her on.

What could she do? Her stomach was churning. Her mind would not settle. It darted like a fly, first one way and then the other. She ached all over. Rest. That was what she needed. Or was it what Fire needed? Their minds had become so intertwined during the extreme stress of the recent battle that Nolita was finding it difficult to tell where the dragon's thoughts finished and hers began.

Fire was exhausted. There was no denying that. She, too, was aching and drained.

Tears streamed down her cheeks. It was so unfair. The questors had come so far together, yet in a single day they had been scattered. Where was Pell? He was supposed to be with her. He was a pain when he was around, but even his arrogant smirks and foolish posturing would be welcome right now.

For the first time since Shadow and Pell had separated from them, Nolita began to think through the possible consequences of their actions. She had been so caught up with her own fears, she had not considered what might have happened to Pell and Shadow.

Fire had defeated Deepshade by using a tactic from another world. Shadow was a strong dragon, but what if she had been outmanoeuvred? To the night dragons, Pell and Shadow were outlaws. If they had been overcome, then they would be shown no mercy.

'They could be dead.' She whispered the thought aloud.

'Do not dwell on that thought,' Fire said suddenly, his voice echoing his weariness. *'Shadow is resourceful. She will survive.'*

For a moment, Nolita turned her fingernails into claws and pressed them through her hair and into her scalp. Although well meant, Firestorm's intrusion on her thoughts was unwelcome. The last thing she wanted right now was a reminder of his presence. However, having heard his comment, she could not ignore his words.

Logic told her that Fire's assurance was wishful thinking. There was no rational way that Fire could be sure of Shadow's survival. With the possible exception of Widewing, Shadow was the biggest, scariest dragon Nolita had ever seen. If Shadow had been facing any other single dragon, Nolita would not have worried. But she wasn't. Big as Shadow was, the chances of her prevailing against two attackers of similar size and ferocity were not good. Besides, if Shadow had survived, where was she now?

After their fight with the night dragon, Fire had insisted on flying as far as he could. Despite his fatigue they had pressed on for a long time, threading through the valleys until they reached the western-most slopes at the far side of the mountains. If Shadow were alive, would she find them here? It seemed unlikely.

'Our best chance of meeting up with Shadow is to keep heading for the day dragon enclave,' Fire told her, clearly still following her thoughts. 'She and Pell know where we're heading. Their goal is the same. Without the strength of the enclave, we will have no chance of getting to the Oracle and Segun will win. Whatever Pell and Shadow might think of the quest, thwarting Segun will outweigh all other priorities. We will see them again. I am confident of that.'

Nolita wanted to tell Fire to be quiet: to stay out of her mind. But she could not bring herself to do it. Unwanted though his intrusion was, she could not deny his argument. Pell hated Segun. Hatred was a strong motivator. He would follow the plan through. But even assuming he and Shadow had survived, this did not help her present predicament. She was alone with her dragon. Her deepest fears had sprung out of the dark corners of her mind. She was quivering uncontrollably and her stomach was churning. She needed a way to function – a way to overcome the fear.

'You have already developed a successful coping strategy,' Fire assured her. 'Make a shelter. Light a fire.

175

Follow your after-landing routine and get some rest. In the morning get ready as if Pell and Shadow were here. You don't normally talk with Pell when you break your fast. Imagine him eating in silence alongside you. Think on it. But get your rest. We should leave early. The enclave will need some time to muster. We must get word to them as fast as possible.'

Fire always seemed to have an answer. She wanted to thank him, but her mind refused to reach out. The bond represented one of her biggest demons and she could not bring herself to touch it. Not at the moment. It was too difficult. Fire understood. Even within her own mind, she could feel his empathy.

Uncoiling her body slowly and wiping the tears from her cheeks, she realised Firestorm was right. Ritual and routine were her best defence. The crowded trees did not make for an ideal campsite, but the enclosed feel of the forest lent a sense of security. She drew a deep shuddering breath and gathered her focus. The first job would be to find suitable materials for a roof.

'It seems we underestimated Firestorm.'

Pell was surprised to hear the grudging respect in Shadow's voice. *'In what way?'* he asked.

'Look down to your left,' she directed.

Pell looked. Deepshade was immediately obvious

176

on the ground below, and she was a sorry sight. She was squatting with one wing extended and her head was turned back on the arc of her long neck over her injured wing. There was no mistaking the gaping hole that had been blasted through the leathery membrane. Pell pictured Firestorm's burning breath playing over Shadow's shoulder after her injury in Isaa and he winced at the memory.

'*How?*' he asked.

'*I don't know,*' she admitted. '*But it seems we were not the only ones to beat the odds today. The question is "Where are they now?" I'm guessing he and Nolita will likely fly as far away from here as possible before resting.*'

'*Firestorm was almost dropping out of the sky with fatigue,*' Pell said. '*He can't have gone far, can he? Can we catch up with them?*'

'*I would not have thought Firestorm capable of downing a night dragon,*' Shadow replied wearily. '*And certainly not in the state he was in. It seems he is more resourceful than I gave him credit for. Who knows how far he will push before resting? Our friends from the other enclaves are full of surprises.*'

'*So what do we do now? Can you find them?*'

'*They could be anywhere, Pell,*' Shadow said. '*I think our best course of action is to fly westwards and keep our eyes open. Given that we're heading in the same direction, the chances of meeting up are reasonable, but it* is a big sky. If we don't find them, then we will have to continue to the day dragon enclave by ourselves.'

'We should at least make an effort to find them,' Pell insisted. 'I promised Elian we would look after Nolita.'

'And we have done so to the best of our ability,' Shadow said firmly. 'I will try to locate them as we go, but given the evidence below, I would say Firestorm and Nolita are more than capable of looking after themselves.'

Pell could not deny that the discovery of the injured night dragon was a major surprise. He had expected to find the broken bodies of the day dragon and his rider. He was relieved they had escaped, but his promise to Elian was still fresh in his mind. While Firestorm might be fine, there were no guarantees that Nolita was. Being alone with her dragon might be too much for her. The events of today would not have helped her confidence.

'West it is, then,' he announced. 'How far should we go today? I can feel how tired you are.'

'Not far,' she said. 'Let's get out of this valley. Now we know there's no danger of further pursuit, we just need to get clear of the immediate vicinity. We'll make an early start tomorrow and try to catch up with Firestorm in the morning.'

*

178

Firestorm remained absolutely still as Nolita approached him. The day had dawned clear and warm. Nolita had woken instinctively as the sky lightened, even though the sun's golden face would not clear the mountaintops to the east for some time yet. Immersing herself in her ritual preparation, she had tried to shield her mind with the morning routine.

Hands stinging from the vigorous scrubbing she had given them, and her wet hair cold against the back of her neck, she unpacked and repacked her saddlebags. The soothing pattern of organising her belongings helped calm her mind, but eating breakfast proved impossible. Just the smell made her stomach turn. Although she managed to put one small piece in her mouth, she could not swallow. All she managed was a few sips of water.

Despite her efforts to think about anything except flying, she could not escape the inevitable. Stamping out her fire, and smothering the embers with loose earth, Nolita rose and picked up her bags. With her focus on where she was placing her feet, she walked out from under the trees and up to Fire's foreleg. She took a deep breath as she felt her heart accelerating. The familiar black wall loomed.

'Routine. It's just a part of the routine,' she muttered. 'You can do this, Nolita. You *can* do it.'

Fire said nothing. He was like stone. Nolita reached out to place her hand on the same spot she used to aid her first step up his leg every morning. The familiar shock of physical contact with her dragon was particularly intense, causing her to gasp and tense her stomach muscles. But she did not let go. Tears formed at the corners of her eyes. She felt frozen in place.

'Climb!' she ordered herself.

It took a gargantuan effort, but with trembling limbs she climbed first onto Fire's foreleg and then up into the saddle. Getting settled took longer than usual. Her fingers fumbled the buckles on the saddlebags and she struggled to set her feet into the stirrups. But it was not long before the buckles were fastened and Nolita was ready to go. Still Firestorm did not move and she had not felt his touch on her mind through the bond since she had woken at first light.

'Let's go, Fire,' she ordered aloud.

He rose to his feet and, much like in the Chamber of the Sun's Steps the first time she had ridden him, she sensed his joy as he launched into a run and extended his wings. That first solo flight had been momentous, but this one was equally so. Today she had come of age. For the first time since the quest began, Nolita really *felt* like a dragonrider.

Chapter Sixten
'Will You Help?'

'Before we go any further, Jack, I want to see this dragon of his,' Squadron Leader White interrupted, his expression unreadable. 'Pardon my scepticism, young man, but I can't help thinking this is all some sort of jolly jape at my expense. If it weren't for the Top Secret communiqués from Whitehall, I'd be back in the Mess room now with the chaps, but the Generals don't play jokes with Top Secret typed on them.'

'Top idea, Boss,' Jack replied. 'Is that OK with you, Elian?'

'I suppose so,' he said reluctantly. '*Warn Fang and Kira,*' he added silently to Aurora. '*I'm bringing company. I don't like the way this is going, Ra. What do you think?*'

'*Jack is trading his information,*' Aurora replied.

'Bartering is common in many cultures. I sense he truly believes he has what we need and that the exchange is a fair one, but he is holding something back. Be careful. Don't agree to anything yet. It will be easier for me to read his surface thoughts when he is closer.'

'Thanks, Ra.'

'If it helps, there is no feeling of a lie in what he says. His purpose is to capture a single man. My dragonsense is prickling. There is something of destiny in his request. The other man is more difficult to read. All I can tell with any certainty is that he does not yet believe we dragons exist. It will be interesting to see if he becomes any more open when he meets us. Let's listen carefully to what they have to say.'

The two airmen retrieved overcoats and strange, flat-topped hats with a semi-circular brim that jutted out above their eyes. Elian was intrigued and looked closely at them. Both sported badges on a band of material that encircled the hat just above the front brim, though they were different in design. One had a circle of leaves with some interlocking symbols in the middle and the other had a crown above wings, with something else underneath.

They left through the front door, crossing the courtyard and road without seeing anyone. Elian did not think the two men would want to crawl through the hedge in their fine clothes, so he asked if there

182

was a preferred way into the field. Jack led them down the dark lane for some distance until they reached a wooden gate. Elian's eyes were still struggling to adapt to the dark. He did not see how Jack unlatched it, but within a moment they were able to walk through. Jack paused a moment to secure it behind them.

The moon peeped out from behind the clouds, bathing them in its gentle light. The silvery illumination made it easy to find their footing as they walked across the field, but once under the leafy canopy of the trees the darkness closed in, and only the faintest dapple of moonlight filtered through to ground level.

Despite the growing pain in his leg, once cloaked in the deeper darkness Elian felt more relaxed. The fire burning in his wound was spreading steadily in his right thigh and up as far as his hip. Every step was bought with a penalty of pain.

Elian could feel Aurora not far ahead. He led the way between the trees carefully, using branches and trunks for support whenever possible. He did not try to move silently, as he wanted to give Kira warning of their approach. A tiny flicker of light ahead pinpointed his companion's position. Kira had built a covert fire and was heating water in a pan to brew a hot drink. She stood up as they approached.

'Kira, this is Jack,' Elian began.

'Hello again,' she said. 'It's nice to meet you without the air being full of stingers.'

'Stingers? Oh, you mean bullets!' Jack said. 'Yes, things were rather too hot for a proper introduction last time. It's a pleasure to meet you in more civilised circumstances. I must thank you for your help that day. I doubt I would be here if your dragon hadn't shielded my aircraft from the enemy fire.'

'Fang says to tell you, your thanks are accepted.'

'And this is Jack's overlord, Squadron Leader White,' Elian continued.

'Overlord!' White choked. 'No, no! It's not like that. I'm his C.O. . . . his Commanding Officer. I'm not a dictator.'

'That's a matter of opinion, Boss,' Jack said, unable to contain a chuckle at his C.O.'s discomfort.

'Very funny!' White said. 'It's a pleasure to meet you, young lady, even if I can't see you properly. It's a bit dark here. I was told you flew here on dragons. I know my eyes haven't had much of a chance to adapt during the short walk, but I don't see any.'

Elian smiled. Aurora was right. His voice held an undertone that was mildly mocking. It was time to educate the Squadron Leader.

'Aurora, would you mind giving us a little extra

light, please?' he asked aloud. 'Keep it faint, though,' he added quickly. 'I don't want to alert the entire neighbourhood to our presence.'

'*Certainly, Elian.*'

Suddenly, a line of shadow behind Elian that White had taken to be bushes began to glow with a faint golden light and a huge, unmistakeable outline formed. Squadron Leader White's face paled to the colour of his name and his eyes went wide.

'Good God!' he exclaimed. 'Is that real?'

Aurora lifted her head smoothly from where she had been resting on the ground and swung around on her long neck to regard him with both eyes. She snorted and White took an involuntary step backwards.

'I'll take that as a yes,' he said, his voice sounding tight with surprise. 'And there is another one of these creatures?'

'Right next you,' Kira replied casually. 'Why don't you say hello to him? His name is Longfang, but he prefers to be called Fang.'

The Squadron Leader looked around rapidly. 'What? But I don't see anyth—' he began, at which point Fang seemingly materialised from thin air alongside him. 'Holy mother of—! Where did that one come from?'

'*He* was there all along,' Kira told him. 'Fang can

camouflage himself. He becomes all but invisible when he chooses.'

'So it's all true!' White breathed softly, looking up at the enormous shapes.

'Every word, Boss,' Jack confirmed.

'Well I'll be a monkey's uncle! If that isn't the damndest thing!' White exclaimed, stepping slowly away from both Aurora and Fang. He was clearly having trouble believing his eyes. Aurora dimmed her inner light until she became a black shadow again. 'If I'm honest, Jack,' he continued. 'When I first read your file, I thought you must be completely crackers. Now I'm not sure if you're crazy, I'm crazy, or we're both completely nuts. And do you know what the maddest thing is? Having seen these creatures, I suddenly think you might just have a chance of pulling this off.'

'Pulling what off?' Kira asked.

'Jack wants to trade,' Elian explained. 'He says he has information about the final orb, but he wants help with something in exchange for the information.'

'Actually,' Jack interrupted. 'I haven't told Elian this yet, but it's specifically *your* help I need, Kira. There is an enemy pilot nicknamed "The Red Baron", because of the distinctive red triplane he flies. I've flown on Aurora, so I know your dragons can carry more than one person. I want you to take me on Fang's back and—'

'Now wait just a minute . . .' Elian interrupted.

'Let him finish, Elian,' Kira said quickly. 'He's talking to me and I want to hear what he has to say.'

'Thanks, Kira,' Jack said gratefully. 'I need your help to capture him. It has to be you because of Fang's ability to become invisible. Aurora would attract too much attention. We need to keep our mission a closely guarded secret. If the enemy realise we're hunting him, they might try to shield him, or move him to another area.'

'Why this man?' Kira asked. 'Why's he so important?'

'He's their top flying ace,' White explained, a note of passion in his voice for the first time since Elian had met him. 'And he's a menace! He's shot down more than seventy of our aircraft – about a dozen of them in the past fortnight alone. He's slippery as an eel basted in grease. One of our best pilots, Billy Bishop, came close to taking him down last year, but von Richthofen even gave him the slip. If we could take the Baron prisoner, it would dent the enemy's morale and give a huge boost to our chaps.'

'Will you help?' Jack asked.

'It isn't fair for you to ask her that,' Elian said, his anger barely under control. 'Finding the final orb is *my* task. It should be me and Aurora taking the risks, not Kira and Fang. They've suffered enough for this quest. You've no idea what they've been through.'

'I'll need to think about it,' Kira replied, ignoring Elian's outburst. 'I want to talk with both Elian and Fang before I decide.'

'That's fair enough,' Jack said, maintaining his focus on her and ignoring Elian. 'I won't pretend there's no danger. There is. You're right to take your time.' He turned to face the Squadron Leader. 'We'd better take steps to seal off this wood, Boss. We don't want word getting out about the dragons. Can we draft in more men to help patrol the perimeter?'

'I'll see to it,' White replied. 'Come on. We'll head back to the mess. I'll wire HQ for extra bodies. We need to move fast on this one. Goodnight, Elian. Goodnight, Kira. I'll be leading the dawn patrol tomorrow, but I'll send Jack to see you first thing in the morning ... unless you need accommodation tonight?'

'We'll be fine out here with the dragons,' Kira replied. 'We've got used to camping during the past few weeks. And it will give us a chance to talk.'

'Very well. Until tomorrow then.'

Jack and his Squadron Commander left, picking their way through the trees in the direction of their base; White kept glancing back every so often, shaking his head in disbelief. Elian sat down and gently massaged the area around the wound on his leg. It felt swollen. As he sat in silence, he could hear the excited whispering of the airmen for some time as they retreated. When he was sure they were out of earshot, he addressed Kira.

'What are you thinking, Kira? Are you seriously considering taking Jack on this manhunt of his?'

'Do we have any choice?' she asked. 'I don't think we do. Fang says he's willing. He and Aurora both believe Jack has the information we need to lead us to the orb. He also says his dragonsense is prickling. I'm not sure if that's good, or bad.'

'It could be either,' Elian said, his tone glum. 'Aurora said the same thing to me earlier. She's mentioned it before. It seems to be like a sixth sense attuned to a dragon's life purpose, but it doesn't guarantee a happy outcome.'

'Fang explained that, but I can't help thinking it's another sign telling me this is meant to happen.'

'But it should be *my* task,' Elian insisted. 'You shouldn't be doing all the dangerous stuff.'

'I'm not getting the orb,' Kira pointed out. 'I'll just be getting the information you need to find it. Believe me, I don't want to actually lay hands on another orb. I don't pretend to understand the

189

riddling words of the final verse of the Oracle's rhyme, but when we find the orb, you can be sure there'll be something unpleasant waiting for you to deal with. You're welcome to it.'

'Thanks!'

'You know what I mean.'

'I do,' he said, his tone mellowing. 'And I also appreciate your willingness to help. Your hunting skills and instincts make you better equipped for Jack's mission than me. Perhaps it's for the best.'

Elian shifted his position, trying to get comfortable. A particularly nasty spike of pain shot from his leg and ran all the way up his back, making him gasp involuntarily.

'Are you all right?' Kira asked, kneeling at his side and placing a hand on his shoulder with a light touch.

'Yes and no,' Elian replied. 'I'll be fine with a bit of rest, but the walk seems to have inflamed my wound more than I expected.'

'Lie down and get some rest, then,' she ordered in a firm voice. 'I'll get your blankets for you. The weather looks as if it will hold fair tonight. We'll get the dragons to shelter us and we'll build a shelter tomorrow if it looks like we're going to be here another night.'

'You think you'll be able to find this "Red Baron"

in a single day?' Elian asked after her, raising his voice as she walked away to retrieve his saddlebags from Aurora.

'I don't know until we try,' she called back over her shoulder. She returned quickly with the bags and began unbuckling them with nimble fingers. She felt first in one bag and then the other before drawing out his blankets. In the dark there was always the danger of inadvertently pulling something else out with them. Anything small could easily be lost amongst the leafy mulch under the trees. 'We'll do our best to find him quickly,' she continued. 'I don't want to stay in this world any longer than we need to.'

'I know what you mean,' Elian said, the constant grumbling of alien warfare sounding suddenly closer. 'There's something very wrong with a world where people feel the need to develop weapons that can kill at such extreme range. It's bad enough that people fight and kill at all, but the weapons here separate the fighter from the consequences of his actions. Perhaps if the soldiers had to look their enemy in the eye as they fought, they might be quicker to resolve their differences.'

'I'm not sure I agree,' Kira said thoughtfully, spreading the blankets over his legs. People will always fight for a cause they believe in, no matter how bloody the outcome. In a war on this scale the leaders are the only ones who can forge peace. If the soldiers are distanced from their enemies, how much further will their leaders be from one another? I'm just glad that in Areth there are no Overlords with enough men to wage war on this scale.'

Elian fell silent. He could find no comfortable answer to her reasoning. The thought of a war like this in Areth did not bear thinking about, but once the image had formed in his mind it was hard to dispel. Wrapping his blankets tightly around his body he lay back and stared up into the treetops. Long after Aurora had curled around him and covered him with her wing, he continued staring and thinking. By the time he finally slipped into an uneasy sleep, dawn was already threatening.

Chapter Seventen

Manhunt

To Elian it seemed he had just drifted into sleep when Kira woke him with a gentle shake. It was barely light enough to be called morning.

'Jack's coming,' she said, turning to the little fire she had clearly been coaxing back into life. 'Fang and I have decided to take him on his manhunt.'

'Do I get a say in this?'

'No.'

'Then it's fine by me,' he said.

He groaned as he tried to sit up. His leg was pulsating with pain. The bandaging felt too tight. He put his hands on his leg with the intention of giving it a gentle massage, but pulled them back instantly as if burned.

'Ow!'

'Leg hurting again?'

'You could say that.'

He shuffled forwards on his bottom until he could push himself upright, taking all his weight on his left leg and hopping until he had established his balance. Dabbing at the ground with his right foot sent shockwaves of pain through his thigh and brought tears to the corners of his eyes.

'That bad?' Kira noted, getting to her feet and lending him a shoulder to lean on as he hopped across to a fallen tree trunk. Taking his weight on his arms, he lowered himself gently into a sitting position, wincing as his full weight settled on his bottom. Kira returned to the fire.

'I must have overdone it yesterday,' he replied, trying to sound as normal as possible. 'It feels swollen again. I'm going to have to release the bandages soon, or my leg might explode.'

'Do you need a hand?'

'I'll manage,' he said, his face flushing. 'I'll do it after you've gone.'

'Your choice,' she said, not looking up from feeding small sticks to the hungry tongues of flame. 'I've seen plenty of boys' legs before. In my village, children often went around naked in the summer months until they had seen about ten season rotations. The older ones didn't wear much more. I don't get embarrassed easily.'

Elian was lost for words, but was saved from trying to answer by the sound of approaching feet. It was Jack. Elian hoped the airman had not heard the conversation.

'Morning, Kira. Morning, Elian. What did you decide?'

'I'll take you flying,' Kira answered. 'But I want to know your plan first. Trying to pluck him from his machine won't work. He'll have straps holding him in his seat.'

'My plan is to have Fang grab von Richthofen's machine from above and force it down into allied territory,' Jack suggested.

'*What do you think, Fang?*' she asked. '*Could you do that?*'

'*I don't know,*' Fang answered. '*I could try, but the machines have little substance to them. If I grab the machine and it falls apart, then the man we're looking to capture will die.*'

'Fang's not convinced,' she relayed. 'He thinks your flying machines are too fragile.'

'Damn and blast!' Jack cursed. 'He might be right, at that. But we must try. I'd rather not kill von Richthofen, but if there's no alternative, then his death will be acceptable.'

'To you, maybe,' Kira replied, her voice flat and cold. 'But it's not acceptable to me. Fang is not a

195

killer for hire, no matter what the reward. What other options do we have? Can we snatch him from the ground?'

'I suppose it's possible.' Jack scratched his chin as he thought about it. 'Our intelligence sources have given us a pretty comprehensive briefing. Here, look at this map.'

Jack pulled a folded piece of parchment from his pocket that was unlike any that Kira or Elian had ever seen before. It was thinner and whiter than the parchment used for books in Areth. When he unfolded it, they were amazed to see an incredibly detailed map had been drawn on one side. Every symbol was precisely drawn.

'In a nutshell, Richthofen and his Jasta – sorry, his team of pilots – are based somewhere between the town of Peronne, here . . . and this town, Cerisy.' Jack pointed at the places on the map. 'This line here is the Somme River. The whole area is a mess. You'd need to be harbouring a death wish to try to fly there deliberately in one of our planes . . . but on an invisible dragon? It could be worth a shot.'

'That doesn't sound clever, Kira,' Elian said.

'How do we get there from here?' she asked, ignoring the warning in Elian's voice.

'We're here,' Jack said, pointing to a blank area on

the map. 'We fly north until we reach the river and then follow it.'

'If these dots are towns, that's a large area,' Elian observed. 'How will you find him amongst the thousands of men out there? Do you even know what he looks like?'

Jack reached into his pocket again and pulled out a picture. The image was remarkable. Elian and Kira had never seen such a lifelike portrait.

'One of our spies brought this back from Germany,' he said. 'The Red Baron is a national hero. Getting a picture of him was easy.'

'The artist has captured his expression in amazing detail,' Kira observed, studying the picture of the young man carefully. 'I've never seen such a clever piece of art. But Elian's right, even with a painting like this we'll struggle to pick him out while on dragonback.'

'It's a photograph, not a painting,' Jack said. 'I can't explain photography, but the process produces fine pictures ... and you're right – it'll be a needle-in-a-haystack job. We'd do better to concentrate on looking for his machine. A red triplane will be easier to spot than a man's face.'

Kira took a final look at the face of the man they were to hunt. Straightening suddenly, she pointed at Fang. 'Mount up, Jack,' she ordered. 'We'd better get going. We're not going to find him by standing here and talking about it. Let's go for an initial scouting trip and see what we can find.' She buttoned up the front of her jacket and pulled on her mitts.

Jack followed her example. He fastened his jacket, put on a pair of gloves and a soft leather cap that covered the top, back and sides of his head, extending down over his ears and fastening under his chin. A pair of transparent eye covers, attached around his head by a stretchy band, completed his outfit.

'Those eye covers are such a good idea!' Elian commented, admiring Jack's goggles.

'I'm amazed you dragonriders can fly without goggles,' he replied. 'It must be almost impossible to see anything. I'll get you both a set of headgear later.'

'Be careful,' Elian warned Kira.

'Always,' she replied.

Fang dropped his camouflage long enough for Kira and Jack to climb onto his back and then disappeared from beneath them as if he had never been there at all. Elian could not help but stare at the impossible sight of Kira floating through the air between the trees with Jack immediately behind her.

'And good luck,' he muttered under his breath.

*

'How should I hold on?' Jack asked as they reached the edge of the trees.

'You can hold onto my waist,' Kira told him. 'But if we don't find the man this morning, then I'd like to get some handles stitched to the back of my saddle for you this afternoon. Bring something soft to sit on, too – a blanket, or something. Riding a dragon is uncomfortable to begin with. Unless we get lucky this morning, you're going to be sore by this afternoon.'

'Thanks,' he said. 'I'll bear that in mind.'

'Ready? Here we go.'

From a standing start by the edge of the trees, Fang accelerated to flying speed within a dozen paces, the first downstroke of his wings skipping them briefly from the ground and his second carrying them into the air. Kira heard the gasp from behind her and grinned. Jack's grip tightened around her waist during the acceleration, but it was not long before the pressure from his hands loosened again. She could not imagine many people adapting to the experience of flying on an invisible ride so quickly.

'Which way?' she asked over her shoulder.

'Turn right until the sun is at our back-right quarter,' he replied. 'Then keep going until we reach the river.'

Fang turned and they climbed to a comfortable height, heading north. The distant ribbon of muddy water soon became apparent and the hunt began. It was not long before they spotted their first formation of aircraft, but as they got closer they identified it as a friendly patrol.

For the next three hours Fang flew up and down the battle lines, crossing back and forth between friendly and enemy territory. In some places the lines became blurred. Jack pointed out some of the towns as they went, explaining which ones had fallen to the enemy in recent days. They could see fierce fighting on the ground, but in the sky the activity was not so frantic. They did see some enemy aircraft and flew close enough to determine that none of the machines were the distinctive red of the one they were looking for.

About an hour before midday they witnessed a clash between formations of Allied and German aircraft. Kira sensed Jack tensing behind her, as they watched the wheeling fight break out.

'*He is itching to join the fray,*' Fang told her.

'*He can itch all he likes,*' she replied. '*We're not going to scratch it for him unless The Red Baron is there. Are any of the machines painted red?*'

Fang turned his head to fully scan the fight with his good eye. '*No,*' he informed her. '*Some of the*

*aircraft have a red band around the front, but those have
the friendly markings on their wings.'*

Shortly after midday they headed back to the
wood. Aurora was asleep, her long body curved
around several trees. Elian pushed himself up from
where he had been waiting anxiously next to the fire.
Kira was impressed to see that he had fashioned
himself a pair of crutches.

'Elian mentioned his wound the other day. Is it
bad?' Jack asked her discreetly, as they dismounted.

'He had a piece of metal from one of your flash
bang weapons deep in his thigh,' she explained. 'It
has been cut out, but he's overusing his leg and it's
swelling again.'

'Shrapnel wounds are often nasty,' Jack commented softly. 'I noticed him limp a bit last night,
but had no idea he was suffering. We've got medics
nearby who can look at it for him.'

'With rest his wound will heal fine,' she assured
him. 'He insisted on coming to find you last night
when he should have let me do it. He can be
stubborn like that.'

'Good lad!' Jack breathed approvingly. 'The best
sort.'

'How did you get on?' Elian asked. 'Any sign?'

Kira shook her head and stretched her back. She
was stiff. It had been over a week since she had flown

201

for that long. She looked at Jack to see how he had fared, and it was all that she could do to keep from laughing. He was standing awkwardly, wearing a pained expression on his face.

'Would you like us to cook you something before we go again, Jack?' she asked. 'Or would you prefer to walk the kinks out of your legs and get food from your own people?'

Jack grimaced. 'I think I need the walk,' he admitted. 'But why don't you join me?'

'Oh, no thanks,' she said quickly. 'I'll be fine here with Elian.'

'Kira explained about your injury, Elian.' Jack touched Elian's shoulder gently and looked him in the eyes. 'There's no need to walk far, old boy. I'll have someone bring a vehicle across the field and pick you up from the edge of the wood. Why not come and let someone else cook up some grub for you? And I'm sure we can find you something more comfortable than the ground to sleep on tonight.'

'If you go, I'll come with you,' Kira said. 'You need to eat and rest your leg properly. Aurora will be fine here. There're guards all around the wood. Jack says no one can come in without Squadron Leader White's approval. Aurora will be fine and you should still be able to talk to her from there.'

Elian looked first at Kira and then at Jack. He

was not happy at being cornered, but his resistance crumbled as he thought of all the fascinating things he had seen in the building where he had met Jack and his Commanding Officer.

'All right,' he agreed reluctantly. 'But let's not get too comfortable. I don't want to drag this out a moment longer than necessary.'

Chapter Eighteen
Gathering Forces

'Are you sure this is the right place, Shadow? Where are the guard dragons?' Pell asked scanning the volcano for the telltale blue of a day dragon. 'I don't see them.'

'Neither do I,' she replied, her voice sounding uncharacteristically nervous. 'I have searched the entire area with my mind and my eyes, but it looks like there are none.'

'What? None! Are you sure?'

'As sure as I can be,' Shadow confirmed. 'It appears the day dragon enclave is not concerned enough to mount a guard.'

'If I were a day dragonrider, and I knew Segun was in control of the night dragon enclave, you can bet your life I'd be concerned enough to have someone on watch,' Pell said, amazed. 'What are they thinking? If Segun succeeds and the Oracle dies, this place will be top of his list to visit . . . along with several hundred night dragons. Can you sense Firestorm at all?'

'No,' Shadow replied. 'Unless he's a long way underground, I think it safe to say that he has not arrived yet.'

Pell grimaced. From the moment he and Shadow had been forced to part from Nolita and Firestorm, he had felt sure he would never see them again – at least, not alive. Nolita had been on the edge of losing her self-control since he had first met her and Pell did not believe her capable of functioning as Firestorm's rider without support. The odds were, she had run away the moment they landed. He was not looking forward to explaining what had happened to Elian and Kira.

'What do we do now?' he asked, eyeing the entrance into the mountainside with more than a little nervous anticipation.

'We go in, I suppose,' Shadow answered. 'I'll broadcast a general friendly message as we enter. I don't want the day dragons to get over-excited and toast us on our way in.'

'Good idea,' Pell agreed. 'I'm all for a warm welcome, but not that warm.'

Shadow descended over the lush green of the forest towards the side of the volcano. The air shimmered with moisture above the treetops, but the sky was clear of cloud and it was surprisingly warm. Pell had his jacket open and had flown all morning without wearing his gloves or overtrousers.

As they made their final approach, the enormous cave entrance gaped wide and dark, like a gigantic stone mouth twisted sideways. For a moment Pell pictured gargantuan stone teeth springing from either side of the opening ready to chomp shut on them as they entered.

An old man with white hair and brows awaited them in the huge entrance chamber. Several tunnels led into the volcano's interior. Some were only large enough for men to walk in single file, others were large enough for a dragon to enter with ease.

'Welcome, young man,' the old man called. 'I assume you are Pell, and this fine dragon of yours is Whispering Shadow.'

'That is correct, sir,' Pell answered. 'May I dismount?'

'Please do,' the man offered, reinforcing the invitation with an open handed gesture. 'I've been expecting you. But where are the others? I thought you would have learned to stick together by now.'

Pell slid down Shadow's side and landed lightly. He looked around cautiously before answering. It was hard to shake the feeling that a whole army of day dragons might appear at any moment.

'That is a long tale, sir,' Pell answered carefully. 'I was hoping to find Nolita and Firestorm here. We were pursued and decided to split up some days ago.'

'They have not arrived yet,' the man said. 'But if they were on their way, then I imagine they will get here in due course. I am Barnabas, the senior rider here at our enclave. You do not need to be afraid. I know of your part in the Great Quest. Indeed, I crave news of how you have progressed. Did you secure the Dark Orb?'

'I did, sir . . . Barnabas.'

'Just Barnabas is fine. We do not stand on formality here,' the old man smiled. His blue eyes sparkled under his white brows and his impressive moustache twitched with amusement. 'Then you found the Valley of the Griffins,' he continued. 'Good. Very good. Yet you and your dragon are still intact. If the heart required was not one of yours, then where did it come from?'

'The heart that formed the orb was ripped from Knifetail, a senior night dragon of Segun's inner Council.'

'Really?' Barnabas chuckled. 'I imagine young Segun was not too happy about that. And the Orb of Vision?'

'You seem to know as much about our quest as I do,' Pell observed, his suspicion growing fast.

Accepting Barnabas at face value would be a mistake. The man looked like a favourite old grandfather with his tidy white beard, white hair and twinkling eyes. But Pell knew all too well that appearances could be deceptive. 'How do you know about the Orb of Vision?'

'I solved the riddle poem after your fellow questors left with the first orb,' Barnabas answered with a shrug. 'The Great Quest has been a part of my life for many, many years. I have studied its history for longer than you've been alive. It is hardly surprising that I, of all people, should solve the riddles.'

'And the final orb is . . .?' Pell asked.

'An egg, or an embryo,' Barnabas said thoughtfully. 'I'm not completely certain, but I imagine it to be a dragon egg, for no matter how I twist it around, I cannot see another solution to the first line – "*Life after death from death before life*". Death before life – most would say that life begins when you are born or, in a dragon's case, hatch.'

'But a dragonet is alive in the egg, so how can it be a death before life?' Pell asked, unconvinced.

'You're quite right,' the old rider agreed. 'The rhyme is not technically correct, but it does fit with the nature of the other riddles. Each has required a personal sacrifice of one form or another. What could be a greater sacrifice than that of a life that has not had a chance to really begin? The greatest of orbs is a dragon's device – an egg is the most amazing thing a dragon can make. A life's sacrifice – an unborn dragon for the rebirth of the Oracle – life after death.'

'If you're right, then the chances of success are pretty slim,' Pell observed. 'Dragons don't clutch that often, do they?'

'I wouldn't worry about that,' Barnabas said with a smile. 'If my answer is correct, then there'll be eggs somewhere for you and your colleagues to find. The Oracle would not have initiated the Great Quest unless all the orbs were attainable. Come. Let me get you some refreshments. You have no doubt had an arduous journey.'

'*Is this Barnabas genuine?*' Pell asked Shadow silently. '*He seems too good to be true.*'

'*I sense no word of a lie in what he has said so far,*' Shadow responded. '*He is either telling the truth, or he is extraordinarily good at masking his true intentions. I think you can trust him.*'

'Your dragon will find somewhere comfortable to rest at the end of that passage over there,' Barnabas added, pointing to one of the larger tunnels. 'My dragon, Wiseheart, will meet her and show her where to go.'

Pell nodded and thanked him. '*Wiseheart!*' he exclaimed through the bond to Shadow. '*Tell me he's* joking! If she lives up to her name, these two are some pairing.'

'Riders do tend to take on something of the nature of their dragons over time,' Shadow replied. 'Some say the reverse is true as well, though I'm not so sure about that. Go with Barnabas. I sense no threat in him. I will see you later.'

'Why do I feel like a fly in a spider's web?'

'Relax,' Shadow told him. 'The day dragons and their riders are notoriously noble. Barnabas has invited you to eat with him. According to the day dragon customs that makes you like his brother until sundown.'

'And then?'

Shadow did not answer.

Pell remained unconvinced, but he did as his dragon had suggested and followed Barnabas into what proved to be a maze of tunnels. Every one looked the same. Within a couple of minutes he had totally lost track of where he was and could only hope that Shadow's assessment was right.

Barnabas led him into a large dining hall. The chamber with its long rows of tables was empty as it was neither time for lunch, nor dinner. They sat down at one of them and a serving girl appeared with a large jug of ale and some bread. Barnabas spoke quietly with her for a moment and she bustled away quickly.

'Now,' Barnabas said, placing his elbows on the table in front of him and meshing his fingers together in an arch. He looked into Pell's eyes, his own eyes dancing with anticipation. 'What can I do for you?'

'You're serious, aren't you?' Pell replied, still seeing no deception in the old man's features. 'I don't understand why you're so willing to help a night dragonrider. Day and night dragons have opposed one another for centuries. Why this co-operation?'

'You're not like the other night dragonriders,' Barnabas replied steadily. 'You're different.'

'I don't see how.'

'The Oracle chose you for the Great Quest. That, in itself, makes you different. Come. What is it that you want?'

'I want you to lead your entire enclave to the Oracle's cave and help us deliver the final two orbs.'

'Why on Areth would I need to do that?' Barnabas asked, his eyebrows drawing together into a frown. 'You have the combined resources of every type of dragon amongst your group. What could possibly stop you from . . . unless . . . Segun?'

'The night dragons are blockading the cave,' Pell confirmed. 'When we left Orupee there were upwards of fifty night dragons there with more arriving by the day. I wouldn't be surprised to find

211

the entire enclave waiting for us by the time we return, which is why I need the day dragons to answer in kind. If you can distract the night dragons – draw their attention from the Oracle enough that Kira can slip past and deliver the final two orbs . . .'

'If the day dragons come in large numbers, we risk sparking a conflict between dragons on a scale that has not been seen in millennia,' Barnabas said, shaking his head. 'It could be disastrous.'

'If you don't then, without a miracle, the Oracle will die and Segun will bring that conflict to you anyway,' Pell pointed out. 'He talks about "The age of dragons" that will begin when the Oracle is no more. I'm certain he intends night dragons to control all of Areth before the year's out. I assume you would be opposed to that happening.'

Barnabas clenched his fingers together tightly, his knuckles whitening as he considered his answer carefully. His eyes lost their sparkle and his face seemed to age.

'Dragon has not fought dragon for a very long time,' he said gravely. 'What you're asking carries too great a risk. The Oracle chose the four of you because you have the combined abilities to solve the riddles and overcome the odds to get the orbs to the Oracle in time.'

'You are wrong,' Pell countered, his voice cracking

slightly in his passion to prove his case. 'Dragon *has* fought dragon. Shadow and I have been banished from the night dragon enclave for causing the death of another night dragon. We fought two more, injuring at least one, and possibly both, not five days ago. Firestorm fought another, leaving it stranded with a huge hole burned through its wing. Conflict is coming, Barnabas. Conflict is here. We need your help. I messed up by letting Segun know about the quest. I'm trying to balance that mistake, and I don't know any other way of doing it. I'm here – a night dragonrider on the Great Quest, trying to make sure we don't fail dragonkind and deprive it of a future.'

The serving girl arrived with a steaming plate of food and placed it in front of Pell. Barnabas thanked her absently, lost in thought. The smell of the food rising from the plate was delicious, but Pell did not move. His focus remained fixed on the old man.

'There are one hundred and forty-seven day dragons here in the enclave at the moment,' Barnabas said slowly. 'How many night dragons do you think Segun will use to blockade the Oracle's cave?'

'A lot more than that!' Pell answered fervently. 'Using the Orb of Vision, Kira saw night dragons arriving from all directions, but the main bulk from the enclave had not yet arrived. We can expect to face upwards of three hundred night dragons by the time we get there.'

'The day dragons will be outnumbered by two to one and that is not a good way to start any negotiation,' Barnabas said, his eyes distant and his lips pursed. 'And certainly not good odds if it comes to a fight. Segun is much like his predecessor. If he feels he has the upper hand, he's unlikely to back down.'

'Elian and Kira hope to bring the dusk and dawn dragon enclaves to help.'

'Which might add another fifteen to twenty dragons at best,' Barnabas said, shaking his head. 'That will not significantly change things.'

'We could gather more along the way,' Pell suggested, trying not to sound desperate.

Barnabas regained his focus.

'I hate to admit it, but I fear you're right, young man,' he sighed, drawing a deep breath and straightening his back. 'It seems we can and we must if we are to prevent Segun from destroying the future of dragonkind.' He paused and suddenly looked deep in thought. 'Indeed,' he resumed, 'it seems our numbers are set to increase to one hundred and forty-eight as we speak. Wise has just informed me that Firestorm is approaching from the South.'

'With Nolita?' Pell asked, his eyebrows rising. He

watched as Barnabas communicated the question to his dragon and received the reply.

'Yes,' he said. 'Nolita is with him. Is there any reason why she might not be?'

Chapter Nineteen

Surprises

'What do you think, Husam?' Tembo whispered. 'Is it a rogue?'

'I don't know,' he replied, shaking his head slowly. 'I can't be sure. The villagers say it is, and the dragon doesn't appear to have a rider. That's good enough justification for hunting it. But I've got shivers running up and down my spine. Something's wrong. I'm beginning to wonder if I've lost my nerve.'

The two dragonhunters huddled down behind the bushes again. Tembo took a long, careful look at his friend. Husam was showing no outward sign of fear, but he had always been good at concealing his feelings. His eyes were still their normal shade, which was a relief. Ever since the spirit creature that had possessed Husam was forced from his body at the Castle of Shadows, Tembo had worried that it might try to return. Where had it gone? Had it died when it was driven out of Husam? He hoped so, but something deep down in his gut told him it was still alive somewhere – hunting for a new body to possess.

The night dragon they were currently stalking was a large male that was not showing much sign of intelligence. It had foolishly chosen to rest in a narrow box canyon, which had made the hunters' job easy. The people in the nearby village claimed it had been eating their sheep and cows, but there were no eyewitnesses. All of the evidence was circumstantial. They claimed livestock had been disappearing both during the day and at night, which made Tembo suspicious. Night dragons rarely ate during the daytime and almost exclusively hunted by night. The only exception to this was when a night dragon hunted food for his rider, but Tembo could see no evidence of a rider, so that seemed unlikely in this case.

'Shall we leave?' he asked softly. 'Say the word and we'll go.'

'Yes ... no ... I don't know!' Husam replied, a tormented look in his eyes. 'What do you think?'

'I think there is something very strange going on here,' Tembo whispered, giving a noncommittal shrug. 'The villagers' claims don't add up, but I see no reason why we shouldn't kill the dragon. They're convinced it's a rogue. No one will question the hunt. My thought is that we *need* to do this, Husam. We've already spent nearly all the money we got from selling the spare horses on the new dragonbone weapons. If we don't make a kill soon, we'll have to consider giving up hunting altogether and find a new trade.'

'You're right,' Husam said, his resolve hardening. 'We'll do it. I never was much of a farmer. So, what's the plan?'

Tembo found it strange that Husam was asking him for instructions. Husam had been the natural leader until their visit to the Castle of Shadows. This change of roles was a measure of how deeply his friend had been affected by this recent experience. Tembo thought for a moment. The bodies of night dragons were so heavily armoured that it was impossible to guarantee a mortal wound, even with dragonbone weapons. The most vulnerable area was the torso area just beneath where the wing attached to its body, so their best chance was to attack from the side and strike there, then the dragon would not be able to fly away, and finishing the beast off would become relatively easy.

A sudden keening sound made Tembo plug his

ears with his fingers. The noise was coming from the dragon, but it was unlike anything the big hunter had ever heard before. Keeping his fingers in his ears, he peered around the bushes. What was the creature doing? It had its head tipped upwards like a dog that was howling at the moon, but the dragon's cry was far more piercing than that of any dog.

Tembo turned to Husam. 'Let's do it now,' he urged. 'It's distracted. Try to hit the body near the wing root, then find cover as quick as you can.'

Husam nodded, sucked in a deep breath and got to his feet. With a spear in each hand, the two hunters sprang forwards, sprinting around the bushes. The dragon had nowhere to hide. As one, Husam and Tembo drew back their right arms in readiness to throw. But, just as quickly, they both froze in their tracks as the dragon's keening cut off abruptly and it fixed them with a glittering stare.

There was something about the dragon's eyes that held them – a power. They were not mismatched like Husam's had been when the joining had possessed him, but there was something strange about them. Without warning, the dragon opened its mouth and belched a roaring sheet of fire towards them. The flames burned whiter and hotter than the orange flame of a day dragon. Despite their distance from the beast, a wave of heat raced ahead of the flames that took their breath away. The two dragonhunters did not hesitate. They dropped their weapons, turned, and ran faster than they had ever run before. They passed the bushes where they had hidden, but didn't stop running until they had left the canyon and were hidden deep in the woods beyond.

When they finally paused to rest, Husam looked at Tembo with fear in his eyes. 'What was that?' He panted.

'That was the end,' Tembo replied, his huge chest heaving as he tried to control his breathing. 'Rogue, or not, I'm not hunting a night dragon that can breathe fire. I've never heard of such a creature. Even before the flames came out of its mouth I was scared witless.'

'Me too! There was something horribly wrong with that dragon. It reminded me of Kasau. Can a dragon be possessed?'

'I don't know and I don't want to find out,' Tembo said, shaking his head. 'The villagers only told me the dragon's rider died in an accident, and that it has been terrorising this area ever since. They refused to tell me anything about the rider's death, but I think now that I should have pressed them about it.'

'We should have quit after the castle,' Husam declared, thumping his right fist into his left palm.

'Hunting dragons never used to be like this. They're changing – becoming more dangerous than ever. First day and night dragons start working together and now they're beginning to share abilities that have always been unique to each type.'

'Well, let's be glad we're still alive to quit now,' Tembo replied. He bent forwards, placed his hands on his thighs, hawked and spat. 'I've got the distinct feeling that if we'd thrown our spears, we'd be dead now.'

'Did it talk to you?'

'No. You?'

'No,' Husam said, shaking his head. 'But, like you, I feel as though I've had a lucky escape. Where should we go?'

'Anywhere that's as far away from here as possible!'

'I'm with you all the way.'

'Look, Nolita. Do you see it? We're nearly there.'

Nolita opened her eyes and squinted in the bright sunlight. It took a few heartbeats for her to find focus but, when she did, her stomach tightened with excitement. Fire was right. She could see the two volcanoes ahead. They were approaching the day dragon enclave.

'I see it,' she replied aloud. 'Thanks, Fire.'

For much of the past five days she had flown with her eyes closed, counting Fire's wingbeats – over twenty thousand of them each day. The counting helped to calm her after the trauma she faced each time she had to mount up. It was difficult to believe that her ordeal was nearly over.

Firestorm stopped beating his wings and began to glide in a gentle descent towards the entrance to the enclave. Although the mountains were still several leagues ahead, the way into the enclave was already visible. Would Barnabas be waiting for her like the last time she visited? She doubted it. He had no reason to expect her this time.

'Shadow is already there!'

They were still some distance from the entrance when Firestorm made his surprise announcement. Nolita's chest tightened as she absorbed the news. Her thoughts fluctuated wildly. She was overjoyed at the discovery that Shadow and Pell had somehow escaped the two night dragons but, equally, she was boiling with anger because they had not come and found her afterwards. By the time they landed, she had settled on being angry.

Pell and Barnabas were waiting for her in the entrance cavern as Firestorm landed. Nolita wasted no time. She slid down Fire's side and strode across the short distance to where the two men were waiting.

'Nolita, you've no idea how glad I am to—'

Pell did not get a chance to finish his sentence. Nolita slapped him across the face so hard that the sound of the impact echoed around the cavern.

'That's for leaving me, and this . . .' she pulled her hand back to hit him again, but Pell was too quick. He caught her wrist before she had a chance to land another smack and he pulled her into a hug that left her more confused than ever.

'It's good to see you alive,' he said softly. 'I thought you were dead for sure.'

He released her from the forced hug and stepped back quickly out of range from her quick hands. Nolita was stunned by Pell's embrace. Was his concern genuine?

'*It is,*' Fire told her. '*Shadow is similarly pleased to meet us here. It seems they tried to find us, but when they couldn't, they decided to risk continuing with the plan in the hope of finding us. It was a brave decision for a night dragon and her rider to make. Approaching the day dragon enclave alone could not have been easy.*'

Barnabas smiled, bemused by the strange greeting between the two questors. 'Welcome back to the enclave, Nolita,' he said, stepping forwards and kissing her gently on the cheek. 'It's a pleasure to see you again. Pell has not been here long, but he's told me a little of your adventures. You must be hungry after

223

your long flight. Come. Follow us. We were just sitting down to some food when Wise told me of your approach.'

'Did Pell tell you that Segun is gathering the night dragons to blockade the Oracle's cave?' she asked.

'He did,' Barnabas confirmed. 'Don't worry, Nolita. Your brave journey will not be in vain. The day dragons will do all they can to help.'

Chapter Twenty
The Red Baron

Elian stomped across the room and sat down with a thump on a wooden chair. He stared out of the window into the foggy murk outside, the gloomy weather reflecting his mood.

'This can't go on,' he snapped at Kira. 'We've been here two weeks already and who knows how much longer Jack's mission is going to take! If it weren't for Aurora's ability to control the return gateway and get us back to Areth moments after we left, the Oracle would be dead by now.'

Kira looked at his grumpy expression. She empathised with his frustration. Indeed she shared it, but at least she had been spared the worst by spending most of her time flying, rather than cooped up waiting.

Even though his leg was barely troubling him now,

Elian could not go flying with her. Fang was finding the extra weight of Jack tiring enough. He could not take Elian as well. Kira did not know what to do to speed things up. They had tried using the Orb of Vision to locate their target, but when she looked into the glowing globe, Kira had problems controlling what she saw. It felt almost as if the Orb had a mind of its own and whisked her sight to places that had no bearing on their present situation. After several attempts, they had wrapped it in a cloth and left it with Elian for safe keeping each day.

'We'll have to persuade Jack to tell us what he knows. You and Fang have done your bit.' Elian ranted, looking to Kira for agreement.

But Kira was saved from having to placate Elian by Jack bursting through the door.

'He's back!' Jack announced excitedly. 'The Boss just gave me these reports. The Baron shot down two of our chaps – one just southwest of Bois de Hamel and the other over Vellers Brettoneux. The Boss thinks von Richthofen and his circus are flying out of Cappy, which fits with the other intelligence we've had.'

'Have you looked outside?' Elian said sourly. 'You won't be going anywhere in this.'

'The fog should burn off quickly,' Jack replied, not allowing the weather to dim his enthusiasm. 'The

226

boys from 209 Squadron are planning to launch three flights into that sector in about an hour. If they draw the Baron's Circus across the lines we might get our chance to take him down in our territory.'

Kira saw Elian's eyes brighten as he began to catch Jack's excitement. The airman was already dressed in his flying gear, with his leather skullcap tucked under his left arm and his rifle slung over his right shoulder. After two weeks of scouring the skies over France for the elusive red triplane, it felt good to have a positive lead. During that time there had been much debate on how they could down the Red Baron's machine without killing him in the process. Fang still refused to grab his fragile flying machine, as he was convinced that there was nothing solid enough in its construction for him to gain a good hold on. There was some talk of mounting a machine gun on Fang's back, but this proved too difficult. There was no safe way of securing it and it was far too heavy to hold, so, in the end, it was decided that their best chance of success was to damage his engine.

Jack had insisted that he could use his rifle to take out the engine. 'I've always been a crack shot and these Lee-Enfields are really accurate weapons. I used to take out rabbits with my father's twelve-bore when I was a lad working on the farm. If you get me in close, putting a bullet or two through von Richthofen's engine should be easy – unless he sees us coming. It'll get trickier if he starts manoeuvring hard,' he had explained to Elian and Kira.

'It's not as big as those other weapons you showed us, but it still looks heavy,' Kira had observed. 'Are you sure you'll be able to hold it steady in flight?'

But Jack had thought of that too. The rifle had a sling attached that, when looped around his arm, helped to hold it steady against his shoulder.

In preparation, Kira and Fang had allowed Jack to try out his tactic on an unsuspecting enemy triplane. They had slipped up behind the enemy aircraft and positioned themselves below and slightly offset to the right. Apparently this made it easier for Jack to hold a steady aim with his right-handed stance. True to his word, Jack had put bullet after bullet into the machine's engine. It took three hits before the first signs of smoke appeared, but Jack fired twice more for good measure. Fang confirmed each hit, but the bullets were so small and fast that they were invisible to Kira's eyes.

Even though the wind rush took some of the sound away, and the pilot's leather headgear she wore dulled it further, the crack of the rifle firing just behind Kira's head had caused her to jump with each shot. She hated the noise and did not know how Jack could absorb the brutal recoil of the weapon without appearing in any danger of toppling over Fang's side. He had let her try firing a shot on the ground and her shoulder had hurt for days afterwards.

Because of the thoughtful planning, they were ready and waiting to deal with the Baron and not wanting to waste any more time, now that they seemed to have him in their sights, Kira grabbed her jacket from the peg on the back of the door and struggled into it. Elian's jealousy of her activity was almost palpable, but she did her best to maintain a businesslike bustle and ignore it. Grabbing the leather headgear that Jack had given her, she gave Elian a nod and a weak smile before turning for the door.

'Be careful,' Elian called after her.

'I'm always careful,' Kira called back. 'Don't worry, we'll see you later.'

Kira and Jack walked to the wood in silence. The swirling fog seemed to heighten their anticipation. Aurora was curled up asleep. Despite the murky conditions between the trees, she looked more golden than ever. Resting for the past two weeks seemed to have intensified her colour.

Fang was ready to go, but they had to wait for more than an hour before the spring sunshine thinned the fog enough for him to take off. Kira checked and rechecked the straps on her saddle and pored again over Jack's map. She had memorised many of the key place names on it now. During the past two weeks they had flown across the territory so frequently that she could relate the lines and dots on the map to what the places looked like from the air. Jack had named the places as they flew over them.

Visibility was still poor as Fang began his run across the field, but he reassured Kira that it was safe to get airborne. The sun was a pale-yellow disk in the gloomy sky overhead, and he was content that the fog was definitely going to clear enough for him to land again when he needed to.

Once airborne, they broke through the thinning fog within seconds and the remaining fog was patchy, with clear sky above. The air was fresh, but not cold. Kira felt great with the sun on her face. They climbed rapidly, turning towards the River Somme and straining their eyes for signs of flying machines.

Jack was first to see them.

'Look!' Jack tapped her shoulder and pointed to the northeast, shouting to be heard over the wind rush. 'There. Over Cerisy. I think they're some of our boys, but I can't be sure.'

'*Fang?*' Kira questioned.

'Jack is right,' he confirmed. 'There are two of them. They look like the type of machines he calls "RE8s". But I can also see another much larger group of machines approaching them from further east. Nine altogether. They're of the triple wing variety we're looking for. And I think one of them is red.'

'Right!' she said, surprised to feel the flutterings of fear in her stomach. Or was it excitement? 'Let's go then,' she told him.

'They are already positioning to fight,' Fang warned. 'We won't be able to get there before they start.'

'Do your best, Fang,' Kira replied. 'Remember, it's not our responsibility to stop them fighting. If they want to kill each other, that's up to them. Let's concentrate on doing what we came here for and then get away as fast as possible.'

'Fang thinks he can see the Baron,' Kira called over her shoulder. 'But there's no way we're going to reach him before your friends are attacked. Getting a clean shot at his engine might not be easy.'

'Just give me a chance,' Jack shouted. 'That's all I ask.'

The rhythm of Fang's wingbeats became more urgent as he accelerated towards the distant specks of aircraft, but they were still several miles away when the fight began. Two of the large formation broke away from the main group to attack the

231

friendly RE8s. Fang confirmed that one of them was an all-red triplane. There was no longer any doubt. They had found The Red Baron.

Kira held her breath as the two triplanes dived at the Allied aircraft. It looked like an eerily silent dance of death at this range, though she knew that each side was spitting death at the other. First one, and then the other enemy aircraft broke away. The lead attacker returned to join the main formation, while the other dived and turned eastwards.

'What happened?' she asked Fang. 'Can you see anything at this range?'

'The leader is definitely the one we're looking for,' Fang confirmed. 'The weapons of the RE8s damaged the other pilot's machine.'

'Is the Baron going to attack again?' she asked.

'I think his priorities have changed,' Fang told her. 'Look to your right. He's just spotted more aircraft closing.'

Kira twisted and pointed out the newcomers to Jack. 'Are those machines friendly?' she asked.

'It's the boys from 209,' Jack confirmed. 'It looks like the Baron's Circus have seen them, too. He's leaving the RE8s and marshalling his men for a more dangerous scrap. I don't know if you've seen them, but there's another group of our boys just ahead and below us. The Baron is signalling his men to attack. The sky's getting busy around here!'

Within seconds the sky seemed to become a whirling mess of aircraft turning and diving on one another, with more arriving all the time. Kira didn't know where to look. Just about all the aircraft had red bands around their engines, but she had not yet spotted the one she was looking for.

Suddenly, Jack began jabbing her shoulder again. 'Over there!' he shouted, pointing out to their right. 'Above the main fight – look!'

Kira followed Jack's gaze and spotted the Baron, 'I see him!' she yelled back excitedly. 'What's he doing?'

'Looking out for his junior pilots, probably.'

'*I can sense the Baron's surface thoughts,*' Fang said suddenly. '*He is watching over a relative. The name Wolfram is in his thoughts a lot. Tell Jack to prepare. We'll be in position shortly.*'

Fang turned and the red triplane slid closer and closer. Kira tensed as she felt Jack raise his rifle and settle the sights.

'A bit closer,' she heard him say, more to himself than as a direction for Kira. 'A bit closer.'

CRACK! The sound of the bullet leaving Jack's rifle made Kira jump. No sooner had he fired the shot than the red triplane tipped and entered a steep dive.

'You got him!' Kira exclaimed.

233

'No.'

'No.'

Jack and Fang answered simultaneously, denying the hit.

'I rushed the shot. It went high.'

'It missed the top of his engine by a hand's width,' Fang confirmed.

'Why's he diving then?'

'One of the allied pilots has attacked his relative,' Fang told her. *'The Baron is intent on chasing away the attacker. Hold on tight. I'm not going to let him get away now.'*

'What's that fool doing?' Jack exclaimed as they nosed down into a steep dive to follow the red triplane.

'He's protecting his relative,' Kira shouted back. 'I wouldn't call that foolish.'

'Not the Baron,' Jack corrected. 'The pilot he's chasing. He's wasting all his energy by weaving around. If he flew straight and let his aeroplane get up speed, he'd outrun the Baron easily. His machine's faster than the Baron's triplane.'

The pilot must know who's on his tail, Kira thought. If someone with the Baron's reputation was chasing me, I doubt I'd be thinking rationally, either.

For a moment it looked as though the allied

aircraft might fly straight into the ground. But then at the last possible moment he levelled out, flying at no more than treetop height, and often dropping down lower until he was hopping over hedges and weaving around trees.

Fang bottomed out of his dive at a similar height and it felt to Kira as if time had suddenly accelerated. The world flashed past at a ridiculous speed. They were gaining fast on the red triplane.

'Why hasn't the Baron finished him?' Kira heard Jack wonder aloud. 'He must have had him in his sights a dozen times!'

'His weapons are not working,' Fang said, answering the question that Kira could not help echoing in her mind. *'He's trying to fix them. His surface thoughts are full of frustration.'*

They were racing along the line of the Somme canal towards the town of Vaux-sur-Somme. Kira's heart was racing now. Fang jinked to the right to try to give Jack the angle to fire as they gained.

'The pilot's running at the church!' Jack exclaimed. 'He's so focused on von Richthofen chasing him, he's not looking where he's going!'

'Shoot, Jack, shoot!' Kira urged.

'We're not close enough,' he growled.

'Try!'

The air was remarkably smooth, given the height

they were flying at, but Fang could only match the height of the aircraft he was chasing.

CRACK! Jack fired. Click, click – re-loaded. CRACK! He fired again. No response from the Baron's machine.

'Am I even close?' he gasped.

Kira relayed the question.

'Both shots hit his machine, but neither hit the engine,' Fang replied.

'Hits, but not on the engine,' she called over her shoulder. 'Try again.'

At precisely that moment all hell let loose. Another allied machine overtook them at speed down their left hand side, spraying a torrent of bullets in the direction of the red triplane. Some seemed to find their mark, but none with any more effectiveness than Jack's efforts. The Red Baron turned hard to the right and rolled out heading north-west to climb up Morlancourt ridge. The attacking aircraft was going too fast to follow him. The pilot, having achieved his aim of giving his colleague a chance to escape, instead elected to pull up and enter a gentle left turn over the town.

'Don't lose the Baron!' Jack yelled.

Kira did not have to repeat the message to Fang. He turned so hard and low that his wingtip came dangerously close to catching on the ground.

Crushed against his back, the two riders struggled to maintain their balance. Jack almost dropped his rifle as he fought to stay upright.

Before they had even completed their turn to follow the Baron's triplane, bullets were coming at them from several directions on the ground. Kira crouched low to Fang's back, terrified as the zipping whine of the deadly projectiles filled the air. She felt Fang's pain as several hit him in the chest and legs. More passed upwards through his left wing as he powered up the ridge. Von Richthofen was running for friendly territory but, by doing so, he was crossing the lines at a height that opened him up to a veritable barrage from the ground.

Lying flat against Fang's back, Kira saw Jack out of the corner of her eye. He was not only still upright, but he was lining up for another shot. Was he mad? The Baron's aircraft eased right again, rolling out eastwards. Fang matched the turn, remaining behind and slightly right. The triplane had to be nearly four hundred paces ahead of them. How could Jack possibly hope to score a hit at that range?

CRACK! He loosed the shot and, a second later, the Baron's aircraft pulled up sharply and entered a right-hand turn before nosing down and impacting the ground hard at the top of the ridge.

'You did it!' Kira exclaimed. 'He's landed. Whose

troops are those down there? Look! They're running to take him prisoner.'

'They look like Aussies to me,' Jack replied, his expression unreadable.

'*They will not be taking him prisoner,*' Fang announced, his tone solemn as he eased them away from the ground and entered a gentle climbing turn to the right. '*Jack's last shot hit the pilot in the back. He's dead. I felt him die.*'

Chapter Twenty-One
Jack's Riddle Solution

'You knew!' Kira accused Jack as he slid down Fang's side to land next to her. 'I could see it in your eyes. You knew you'd killed him, didn't you?'

'I suspected I'd hit him, yes,' he admitted with a shrug. 'It wasn't intentional. At that range I was lucky to hit the aircraft. Hitting the pilot was a fluke. As soon as I saw his aircraft pitch up and to the right the way it did, I knew it was likely that I'd injured him – or someone else had. There were a lot of bullets spraying at him from those machine-gun posts on the ground. It might not have been me.'

'No. *You* killed him,' Kira said firmly. 'Fang may only have one eye now, but his vision isn't like ours. He sees things we can't. He tracked your bullet. It went straight in through the Baron's cockpit and into his back. Fang was also listening to the Baron's surface thoughts. He felt the shock of impact. He felt the Baron die.'

'Well, I can't say that I'm sorry,' Jack admitted, jutting his chin defiantly. 'He killed a huge number of our men, including my wingman last year. He was a menace – a brilliant pilot – but a menace to our forces. Don't worry, Kira. He'll receive the recognition he deserves as the worthy foe he has been. I'll get the Boss to see that he's buried with full military honours. April 21st will long be remembered as the day The Red Baron died.'

'He was defenceless and you shot him in the back!' she accused, eyes flashing dangerously.

'Baron von Richthofen has *never* been defenceless,' Jack barked vehemently. 'He was a master strategist, although today he wasn't on form. I don't understand why he made such an uncharacteristically reckless error. He chased his prey – down to low level on the wrong side of the lines, with no support and with limited weaponry. It wasn't like him at all. It's long been known that he is . . . he *was* a careful pilot. I'm sorry, Kira. I didn't mean to kill him. This didn't work out the way I intended. But come. You did as I asked. Let's go and see Elian and I'll fulfil my side of the bargain.'

Kira shook her head with a look of disbelief and stormed off ahead of him. She felt sick and used.

Hunting and killing a creature for food was one thing, but hunting down and killing one specific man felt horribly wrong. Even though she had not personally killed the Baron, she felt responsible. The feeling twisted in her gut until she wanted to throw up.

Elian was momentarily delighted when Aurora gave him the news that Kira and Jack had succeeded in bringing the Baron down in Allied territory.

'It's not all good news,' Aurora continued. 'Jack killed the pilot.'

'Kira and Fang are all right, though, aren't they?' Elian asked quickly.

'Yes ... and no,' Aurora replied cryptically. 'Fang was hit by a large number of stingers, but has not suffered serious harm. Kira was not harmed physically either, but Fang tells me she's deeply troubled by Jack's shooting of the pilot. She feels responsible for the man's death. You will need to tread lightly when she reaches you.'

'I see,' Elian said. It was a reflex answer. He didn't see at all. Why should Kira feel responsible for Jack's actions? He was a soldier involved in a war. Killing had been a part of his life for several years. It was not her fault that Jack had used his weapon against the man, rather than the machine. The mission was a success. She had done what Jack had asked, which brought a pressing question to the forefront of Elian's mind. '*Has he told her the answer to the Oracle's riddle yet?*'

'*No, not yet,*' she answered. '*Jack has promised to tell you both together when they reach you. They are on their way to you now.*'

Elian was out of his seat, through the door and outside in no more than a handful of heartbeats. He could see them coming across the field. Kira was some distance ahead of Jack. Both had their heads down. He could understand Kira's posture, but why did Jack look so depressed? Surely he should be jubilant. He had done what no other flyer had managed. He had shot down his deadliest enemy.

The two figures came closer. When Kira turned from the lane into the courtyard, he ran forwards and drew her into a hug.

'Aurora told me,' he whispered in her ear. 'It's not your fault, Kira.'

Kira did not reply, but she leaned her head into his shoulder and he could feel her weeping gently. Elian was shocked. This was not like the Kira he knew. He never thought he'd see the tough huntress shed a tear, let alone weep. Jack entered the courtyard behind her. His eyes were sad as he met Elian's gaze.

'I'm sorry, Elian,' he apologised. 'I meant no harm to Kira, I swear. I hope she will forgive me eventually. Come. Let's go inside. The Boss needs to wire the news to HQ as soon as possible. I'll give my report, then show you the answer to your riddle poem.'

Jack led the way into the mess building. He left Elian and Kira and went to make his report. From the moment the door closed, Elian began pacing back and forth like a caged cat. Kira sat on a stool and stared blankly out of the window, listening to the sputtering of aircraft engines starting up – the launch of another patrol.

To Elian every second Jack was gone felt like an age. The longer he waited, the more his suspicion of the pilot increased. What was he really doing? Did he actually have an answer to the Oracle's riddle, or had he fooled them into helping him with false promises?

The door opened and Elian turned instantly. Kira remained staring out of the window. Jack entered, carrying a folded map. Why a map? Was the final orb in this world after all? Did he know the orb's location?

'The Boss sends his thanks for your help,' Jack began. 'Apparently there are generals at Head-quarters who want us to use all means possible to convince you to help with another problem . . .'

'But you promised . . .' Elian objected fiercely.

Jack held up his hand, his stern expression halting

243

Elian's outburst. 'If you'll let me finish, Elian,' he continued. 'The Boss's words were, "The Generals can go and take a running jump!". He has authorised me to release the answer to your riddle, but I didn't need his permission. I would have done it anyway. Here. Look.'

Jack moved to the table in the centre of the room and unfolded his map. There were lines of symbols on the back, which Elian knew to be the strange writing of this world. There were five groups of four lines. It had to be the Oracle's riddle rhyme.

'Before I tell you outright what the answer is, would you mind writing out the riddle verses in your own language?' Jack asked. 'I wonder if this works in both of our languages.'

'I'm not too good at writing,' Elian admitted. 'What about you, Kira?'

Kira was still staring out of the window. She did not look round, but she shook her head.

'OK, I'll try,' Elian said.

Jack handed him a writing instrument unlike anything he had seen before. Placing the pointed end against the paper, he was amazed to see ink flow in a neat line as he brushed it against the white surface. Elian wrote the lines out in his own language, doing his best to make sure they lined up with Jack's verses.

When Elian had finished, Jack thanked him and

then carefully underlined the fifth line, the tenth line
and the seventeenth line:

> *Beyond time's bright arrow, life-saving breath,*
> *Love's life force giving, slays final death.*
> *Orbs must be given, four all in all.*
> *Orbs to renew me, stilling death's call.*
>
> *Delve 'neath the surface, life's transport hides,*
> *Healing, restoring – bright river tides.*
> *Enter the sun's steps; shed no more tears.*
> *Attain ye the orb; vanquish the fears.*
>
> *Release the dark orb – death brings me life.*
> *Take brave ones' counsel, 'ware ye the knife.*
> *Exercise caution, stay pure and heed,*
> *Yield unto justice: truth will succeed.*
>
> *Ever protected, the dusk orb lies*
> *Behind the cover, yet no disguise.*
> *Afterlife image, unreal yet real,*
> *Lives in the shadows, waits to reveal.*
>
> *Life after death from death before life,*
> *Enter the new age, through deadly strife.*
> *Greatest of orbs is – dragon's device.*
> *Gifted for ever: life's sacrifice.*

'Do you see anything unusual about the verses now?' he asked.

Elian looked, but he could see nothing. He shook his head.

'The riddles in each verse are clever,' Jack explained. 'If I had been clever enough, then I might have worked them out the way you were trying to do. But, to be honest, I wasn't sharp enough to see the answers. There is an easier solution. In my language the Oracle's poem is what we call an acrostic. You can read words downwards using the first letter of each line. I didn't notice it until I wrote it out on the map. Even then I didn't notice it straight away. Of course, once you see it, the solution is obvious. Try taking the first letter of each line and writing them out across the page like this'

Jack wrote the letters:

BLOODHEARTEYEBALLEGG

Then he added diagonal slashes after the fifth, tenth and seventeenth letters, just as he had underlined those relative lines in the poem.

BLOOD/HEART/EYEBALL/EGG

Elian tried it, but the string of letters were meaningless.

'It doesn't mean anything,' Elian said, confused. 'It's just a string of nonsense.'

'Interesting!' Jack mused. 'It appears the only way the solution becomes obvious is when the poem is written in English. Your Oracle is beginning to scare me. In my language the words read: blood, heart, eyeball, egg.'

Elian went pale.

'"*Life after death from death before life*"!' he gasped. 'An egg! We've got to get a dragon's egg to the Oracle before it hatches! The hatching! But how did the Oracle know about the eggs? That's impossible!'

'*There is little that is impossible for the Oracle,*' Aurora told him, her voice irritatingly pompous.

'*Can we get back before the last egg hatches?*'

'*Of course!*' Aurora answered without hesitation. '*The Oracle would not have given us a task that was impossible. Don't worry. I'll get us back before the last of the eggs hatch. The question is more whether we can transport the egg to the Oracle's cave in time. Once we get back to the enclave I will have to wait a full day until we can form another gateway and it's much too great a distance to fly it direct.*'

COMBATS IN THE AIR

Date: 21-04-18.
Duty: Spl Mission.
Height: Low Level.
Time: 09.45 a.m.
Squadron: 60.

Type and No. of Aeroplane: Invisible dragon!
Armament: Lee-En eld.
Pilot: Un-named civilian.
Observer/Gunner: Capt. J. Miller MC.
Location: NW Cerisy.

Remarks on Hostile machine – Type armament, speed, etc.

All-red Fokker Triplane.

—— Narrative ——

Left field at 09.45 a.m. and at 10.00 a.m
observed formation of 9 E.A. Approaching
Cerisy from the east. Two E.A. (one observed
to be coloured red) engaged two RA8s over
Cerisy, but were driven off.

Two further friendly formations of Camels
arrived in the vicinity at 10.05 and a

dogfight ensued over Le Hamel. The red
triplane I was hunting dived from above the
fight, chasing a Camel acft down into low-
level and along the Somme canal towards the
pontoon at Vaux-sur-Somme. I followed on
dragonback. Scored several hits from mid-range
with my rifle, but none critical.

Further Camel acft attacked E.A. from the left
as we approached the pontoon. Red triplane
broke off pursuit of first Camel acft and
turned right up Morlancourt Ridge. Lots of
ground fire observed from machine-gun
positions on the ground, but my final shot
from long range (behind and slightly below)
was tracked by the dragon and observed as
hitting the pilot in the back. Red triplane
pulled up into a steep turn to the right and
carried out a steep glide landing on the top
of the ridge. Returned 10.50.

I fully understand that this kill cannot be
accredited, as no one will be able to confirm.

J. Miller

BRITISH WAR OFFICE MEMO

21-04-18

Hugo

Please see the attached combat report. It appears Capt. Miller has brought home the bacon! I've been getting confirmations that the pilot of the red triplane was indeed Manfred von Richthofen, but we are now faced with a quandary. What do we tell the press?

There have been several claims submitted. Sqn Ldr Brown of 209 Sqn RAF was the pilot of the attacking Camel acft that Miller mentioned in his report. I'm told he submitted a claim, but then immediately withdrew it. I'm tempted to order his C.O. to get him to re-submit, so we can ratify it with all speed. Since we lost Capt. Ball last year the public have been craving more tales of heroic airmen. This is a great opportunity to give them one. Miller is right in saying we can't credit him. We *must* keep this dragon business under wraps if we ever hope to use them again.

The Aussies have submitted a claim of their own. They say Brown did more damage to their soup kitchen than he did to the Baron's aircraft! Serves them right for being so damned good at cricket, if you ask me.

What do you think?

Maurice

Lieutenant General Maurice Tremelayne

BRITISH WAR OFFICE MEMO

21-04-18

Maurice

Re. your memo earlier today about Capt. Miller - spot on. Get Brown to re-submit ASAP. It will be a great boost to national morale to have another flying hero for them to worship.

We'll have to move fast on this - muddy the waters as much as possible. If anyone starts to suspect things are not what they seem, we'll have the press snooping around in no time.

Hugo

Field Marshall Hugo Fitzpatrick

Chapter Twenty-Two
The Final Orb

'No, Elian! NO!' Tarl snapped. 'You're not taking one of the eggs. I would *never* ask Blaze such a thing. The very idea is monstrous. It's unthinkable!'

'What if I got Aurora to . . .'

'NO!' Tarl's eyes were narrowed, and his face tight with fury. 'Don't you *dare* start Aurora upsetting Blaze over this. If Blaze were human, would you ask her to sacrifice a baby for you? That's what you're asking, Elian. The answer is no, and it will remain no.'

Elian nodded, bowed and withdrew down the corridor with Kira at his side. Neema followed them, her face troubled. They walked in silence until they reached the common room. The oddly shaped chamber was filled with natural light that poured in from the opening to the wide mountain ledge at the far end. Hangings adorned the wall and comfortable chairs, piled with soft cushions, beckoned. Elian threw himself into a chair and slumped forwards with his chin in his hands.

'Well!' he sighed. 'What now?'

To Neema it was less than an hour since Elian had left the dawn dragon enclave. Although she had been a dragonrider for some years, she and Shimmer had never travelled into the other world together. It was clear to Elian that she was struggling to comprehend how he and Kira had lived out more than two weeks of their lives in what had been a matter of minutes for her. He suspected that, without the evidence of how much his leg had healed and Shimmer to confirm they were telling the truth, she might not have believed them.

'It seems your challenge is very different from the previous ones,' Kira said carefully. 'Convincing Tarl and Blaze to part with an egg is not going to be easy.'

'You saw Tarl. You heard what he said,' Elian grumbled. 'He won't listen at all!'

'What you're asking *is* rather a lot,' Neema observed. 'Tarl and Blaze have been preparing for the hatching for over two months. It is a momentous occasion and very unusual. Not many riders get to experience what Tarl is going through. He and Blaze have been utterly focused on giving the hatchling dragons the best chance of survival and here you are asking him to sacrifice one. How would you feel in his place?'

'Not good,' Elian admitted. 'But we're not just talking about the future of one hatchling here. If the Oracle dies, it will be the beginning of the end for all dragonkind.'

'That's what the Oracle has told you, is it?' Neema asked.

'Yes,' Kira confirmed.

Elian just nodded.

'And if the Oracle is lying?' she asked.

'Our dragons are insistent that the Oracle doesn't lie,' Elian answered, though even he could hear the uncertainty in his voice.

'Do you think we could steal one?' Kira asked slowly.

'I'm going to pretend I didn't hear that!' Neema gasped, her hands automatically going to her mouth with the shock. 'For all of our sakes, you should forget that thought ever entered your head. If you tried, Blaze would kill you. And even if, by some chance, you did succeed, she and Tarl would rip Areth apart hunting you down. No. It's not an option.'

Elian interlocked his fingers behind his neck as

he considered what Neema had said. Reluctantly, he was forced to agree with her. Stealing what they needed was not the way ahead. But what choices did that leave?

'*The way will become clear,*' Aurora announced in his mind.

'*What do you mean?*' Elian asked her. '*How will it become clear?*'

'*I don't know exactly, but my dragonsense tells me it will. Trust me.*'

Elian had come to rely a lot on Aurora, but this time her seemingly blind faith that things would work out seemed misplaced. Waiting and watching when there were only a few days left until the harvest full moon seemed a ridiculous thing to do. They needed to get the egg as fast as possible and reach the Oracle's cave before it was too late, or all the pain and sacrifice they had been through would have been for nothing.

'*Excuse me if I don't share your optimism,*' Elian told her. '*I'd like to try to come up with a more practical solution, if I can.*'

He sat for some time, lost in thought. Eventually, Neema announced that she ought to get back and check on Tarl and the hatchlings. Elian knew it was an excuse to escape the awkward silence that had descended since their brief discussion. Once she had gone, Kira broached the subject of a raid again, but Elian dismissed it out of hand.

'Neema was right,' he said. 'And Ra is telling me to be patient – that our path will become clear.'

'What does she mean by that?'

'I'm darned if I know, but I don't think she's keeping information from me,' Elian replied. 'I think she's still expecting an Oracle-inspired miracle. I've got to admit the Oracle does appear to have had great foreknowledge of our quest and the problems we had faced. Maybe Ra's right. Perhaps, just like Jack showing up every time we went into the other world, something amazing will happen to put what we need in our path.'

'Sounds like wishful thinking to me,' Kira muttered.

Elian gave her a wry smile. 'That's almost exactly what I told Ra,' he said. 'But you try telling Fang he's delusional and see where you get!'

Kira began to laugh. It started as a snigger and quickly developed into a full-blown belly laugh. Within seconds Elian found the infectious nature of her laughter too much to resist. His wry smile became a chuckle and, before long, he was laughing along with her.

When he finally managed to stop, Elian realised that that some of the tension had dropped from his shoulders. His frustration had not gone, but suddenly the situation felt less dire. It was hard to remember the last time he had found something to laugh at. He had spent most of the quest racing from one dangerous place to another.

'I'm guessing you don't tell Fang he's delusional that often,' Elian commented as he dried his eyes. 'Perhaps it would do the dragons good to have their failings pointed out every now and then.'

'*My dragonsense is not a failing, Elian,*' Aurora informed him in a matter-of-fact voice. '*Quite the opposite. You'll see.*'

'I'm being put in my place, now,' he said aloud, for Kira's benefit.

'Quite right, too,' she agreed with a final chuckle. 'I don't think Fang would talk to me for a month if I called him delusional.'

Elian walked to the open ledge to look out at the lush green jungle in the valley below. It was barely an hour after dawn, yet the temperature and humidity were rising fast. Steam was rising from the thick, leafy canopy in a mist that gave the place an almost magical aura.

Of the three enclaves he had visited, this was by far Elian's favourite. Was he biased because it was the dawn dragon enclave, he wondered? Perhaps, but there was more to it than that. He felt a connection with this place. The volcanic home of the day dragons in Racafi was impressive, and the enormous high mountain hideaway of the night dragons perhaps even more so. But there was something about the setting of this small group of caves in South Cemaria that felt special.

For the rest of the day, Elian's mood went full circle from frustrated, to relaxed, to angry and back to frustrated again. During the day more eggs had hatched.

'How many are left now, Ra?' he asked as the sun finally set and day turned to dusk.

'Four,' she replied. 'And two of those look imminent.'

'Any sign of your miracle yet?'

'Not yet,' she admitted. 'Give it time.'

'Give it time! Give it time?' He could hardly keep his thoughts coherent as he responded. 'We don't have any time left! When are you going to accept that the miracle isn't coming? We need to do something, Ra. If we don't do it soon, the Great Quest will fail.'

'We are doing something,' she said haughtily. 'We're waiting. Don't underestimate the power of patience.'

'Oh, right!' Elian exclaimed, thumping his fist against his injured thigh. The pain that shot up his back took his breath away for several seconds. 'The virtue of patience will win the day, is that it?' he continued. 'Well I hate to be the bearer of this bad news, Ra, *but the Oracle is unlikely to thank us for waiting if the last egg hatches ... unless you know of another place where we can find a dragon's egg, of course.'*

'I would thank you to keep your tone civil, Elian,' she replied, her tone full of reproof. 'I doubt you would talk to the elders in your village like this, and I'm far older than any of them. Just remember this is my life purpose we are talking about. I've waited centuries for this time. I've spent my entire life preparing; honing my instincts and skills to be ready. Everything I am is telling me to be patient. That is what I'm doing. Do you doubt my commitment?'

'Elian! Kira! Come quickly!'

It was Neema. She looked incredibly excited.

'Ah!' Aurora sighed. 'The waiting is done. My dragonsense is prickling again. I'm surprised you can't feel it.'

'What is it, Neema?' Kira asked, her voice reflecting a hint of Neema's excitement.

'You need to come and see for yourself,' she urged. 'Quickly! Come to the hatching cavern.'

The dawn dragonrider turned and ran back the way she had come. Elian and Kira were quick to follow. When they got to the chamber entrance, they saw Tarl staring wide-eyed through the open door.

'What is it?' Elian asked. 'What's going on?'

Tarl did not answer. He continued to stare in an

260

obvious state of shock. Elian edged forwards until he could follow the line of Tarl's eyes and his breath caught in his throat. There, in the very centre of the chamber, was a metal plinth just like the others they had seen at the location of each of the other orbs.

'Where did that come from?' Kira asked. 'I swear I scoured every hair's breadth of this enclave looking for it and I found no trace.'

'It rose up through the floor,' Neema answered. 'There was no opening – nothing! It just seemed to rise through solid rock.'

'I'm sorry, Ra,' Elian projected. *'You were right. I should have trusted you.'*

'Thank you, Elian,' she replied, her voice not even slightly smug. *'Your apology is accepted.'*

'What are we supposed to do now?' he asked.

'I'm not sure,' she replied. *'But we will find out soon enough.'*

Elian no longer doubted her. As he watched, two of the four remaining eggs hatched, their shells shattering almost simultaneously. The two dragonets protested their abrupt emergence with ear-rending cries, tottering around and each snapping at the other as if blaming it for their circumstances. One of the two remaining eggs wobbled, but the final egg remained unmoving amongst the shattered shards of shell.

'That last one looks smaller than the others,' Kira whispered softly in Elian's ear. 'Do you think—?'

The question went unfinished, but Elian had reached the same conclusion as Kira. The hatchling inside was no longer alive. Death before life – that was how the Oracle had described it. Could it be true? Was it really that easy? After the sacrifices the other teams had been required to make – the blood, the heart and poor Fang's eye, it seemed almost unfair that he should simply be required to pick up an egg and put it onto a plinth. Surely there must be more to it than that!

Taking a step forwards, Elian had not even reached the threshold of the chamber when Blaze's head snapped around, eyes intent and body instantly emitting a warning glow. He stopped and a strong hand grasped his shoulder.

'You cannot go in,' Tarl said, though there was no anger in his voice now – only an echo of his dragon's warning stance. 'Blaze won't permit it. She'll kill you if you go near her remaining eggs.'

'But I've got to go in,' Elian replied. 'It's my destiny to go in there and put that last egg on that plinth. The Oracle requires it.'

'Oracle, or not, Blaze won't listen,' Tarl warned. 'Please, reconsider. I don't want to see you hurt. At least wait until the penultimate egg hatches first.

Blaze isn't in her right mind. The hatching has made her unreasonably defensive. She wouldn't even let me in there to help the hatchlings get to the meat. I sense she already knows there's something different about the final egg, but I wouldn't go in there now if I were you.'

'All right,' Elian conceded, backing away from the entrance. 'I'll wait until the other egg hatches, but no longer. If that plinth disappears again, I'll never forgive myself.'

He did not have long to wait. The final hatchling burst through its shell, gave a single shriek and made straight for the final scraps of meat that the other dragonets had left.

The remaining egg did not move. Blaze slowly lowered her head over it, twisting until the side of her face gently kissed its surface. She remained still like that for some time before slowly rising over it again. Elian looked at Tarl questioningly. The rider's face held a look of tender sorrow.

'She knows,' he breathed.

Blaze opened her mouth then and the sound that emitted was heart-rending. Her keening filled the cavern, flooded out through the door and around the enclave, spilling out of the entrance and across the valley below. It was so piercing that the people of the nearest tribal village two leagues away stopped in their tracks to listen. There was no mistaking the sadness and mourning in her voice, filled with tragedy and loss.

'Tell her,' Elian urged Aurora. 'Tell her it's not the end for her final egg. Tell her I'm going to help it be born into the most amazing being known to dragonkind – that her last little egg is destined to be transformed into one far greater than any of the others. Tell her now. I'm going in.'

'She's not listening, Elian,' Aurora warned. 'Her mind is closed. Be careful. I don't seem to be getting through to her.'

'Keep trying,' Elian said, stepping across the threshold and into the cavern. 'Don't let up.'

'Elian!' Tarl hissed. 'What are you doing? Don't be a fool! Come back.'

He took another step forwards, and Blaze fixed her eyes on him. Her head drew higher and began to sway gently from side to side on her long neck, like a serpent preparing to strike. All the while she continued to keen, her voice rising and falling in a seemingly never-ending cry of sorrow. Elian watched Blaze intently, following the flowing movement of her head as he inched further and further into the cavern.

'I'm not here to harm your final egg,' he whispered, not daring to blink for fear of breaking eye contact. 'I want to help it. Your babies are amazing. You're amazing. All your other eggs have produced fine dragonets. Let me help the last one become something even more incredible.'

A sudden movement in the corner of his eye drew his attention for the briefest of instants. It was the final hatchling. The baby dragon had finished gobbling down the last piece of meat and had begun stumbling across to where all the others lay asleep in a tangled pile. His distraction lasted barely more than the time it took to blink, but Blaze had seen his glance in the direction of her babies. To her mind that one flicker of attention in the wrong direction was enough to make him a threat.

'ELIAN!'

Aurora's shout echoed in his mind, even as Kira's echoed physically in the cave. Blaze lunged forwards at him, her enormous mouth gaping. Elian launched into a rolling dive to the right as the dragon's great jaws drove through the space he had occupied a heartbeat before. In an instant, Elian was on his feet and sprinting again, ducking under the dragon's long neck in a zigzagging run that took him away from the dragonets. Those of the newborns that were still awake hissed in sympathy with the anger of their mother.

Blaze swept her head around in a long, low arc

265

after Elian, seeking to crush him against the wall of the chamber.

'BEHIND YOU!'

Kira's warning was timely, but did not change Elian's tactic.

He was aware of Blaze's move and made another dive, this time flat and to the left. His belly, thighs and elbows grazed against the rock floor of the chamber as he skidded clear. The dawn dragon's head whipped past so close that he felt the swoosh of displaced air. Blaze's cheek hit the chamber wall with a hefty thud, and she let out a roar of pain and frustration.

Scrambling to his feet, Elian took off again, this time in a darting run straight towards the angry Blaze. He could hear Aurora's mental shouts as he ran. She was trying to convince Blaze to stop her attacks, but the mother dragon was paying no attention. Elian's focus was totally fixed on his goal – the egg. He had to reach it. Blaze drew her head back to strike again, but even as she began her downward lunge towards him, Elian covered the final few paces to his goal and wrapped his body around the great orb of mottled shell. Bracing himself for the crushing strike of the dragon, Elian closed his eyes and gritted his teeth.

The expected strike did not come, but he could

feel the warm, damp breath of the dragon against his back. He had gambled on her not wishing to harm her egg even though she knew the hatchling inside was dead. It appeared his reasoning had been sound.

Elian cracked open the eyelid on his left eye and tipped his head to one side. He was rewarded with a terrifying view into the dragon's open throat. Blaze's jaws were spread so wide that her great cage of teeth surrounded him. Even as he began to take in the precarious nature of his position, the dragon's huge forked tongue began to snake under his legs in an effort to dislodge him.

'Oh no you don't!' he muttered, clinging to the egg even harder.

'Don't move, Elian! Stay exactly where you are.' It was Tarl. He had come to help. 'Blaze! Listen to me. Elian doesn't mean any harm. Look at the plinth, Blaze. Your last egg has a great destiny. It is going to restore the Oracle. Isn't that wonderful?'

The tongue withdrew and slowly, the great jaws rose away from Elian. He heaved a sigh of relief and, for a moment, he thought he might pass out. His head spun and a wave of nausea swept through his gut and up into his throat. Tarl had done it. Blaze was listening at last. As the dragon's mouth closed and her long neck lifted her great head higher, Elian began to relax.

Uncoiling his body from around the egg, Elian climbed back to his feet.

'Not yet, Elian! It's too—'

Tarl did not get to finish his warning. Blaze struck, her head arrowing down so fast that Elian did not have time to fully evade the strike. All he could do was twist and duck. The dragon's lower jaw hit him harder than a charging bull, smashing into his right arm and chest and hurling him back across the chamber towards the plinth. Winded, he scrambled behind the minimal cover offered by the small pillar of metal.

To Elian's horror, he saw that Blaze had also swept her own rider aside. Tarl was also picking himself up some distance to the right. But Blaze was not finished. Her eyes were still intent on Elian and she was coming for him again.

With a sweeping motion, Blaze whipped her head down towards him once more, but Elian was hurting too much to do more than cower behind the plinth. The side of Blaze's jaw connected hard with the metal. There was a horrible cracking sound and the dragon's head stopped dead. For a moment Elian had a second terrifyingly close view of the side of her jaw and her left eye before she reared backwards and let out another roar of pain. As she pulled away, one of the large outer teeth from the left side of her mouth fell loose and a drool of dark blood dribbled to the floor. The plinth had not shifted so much as a finger's width.

The dragon lunged a second time and then a third. Each time Elian dodged, keeping the plinth between him and Blaze's anger. Her body was glowing brighter as her anger and frustration built. Elian knew from Aurora's ability that if Blaze threw all her energy into intensifying that inner light, then he would neither be able to look at her, nor see her attacks coming. Fortunately, she was too agitated to use her abilities to the full.

Everyone was shouting – everyone except Elian, who was too busy trying to stay alive. Tarl was bellowing at his dragon. Neema and Kira were yelling warnings to Elian. The roars of Aurora and Fang were also getting nearer as the two dragons approached through the tunnels. What got through to Blaze, Elian did not know, but as abruptly as she had attacked, so she stopped.

Blaze backed off slowly and curled protectively between Elian and her hatchlings, some of which had woken with all the noise and were screeching their high-pitched protests. Her inner light dimmed back to a subtle glow. She eyed Elian suspiciously as she licked around her bleeding mouth with her long forked tongue.

'Tarl? Can I get the egg now?' Elian called, without taking his eyes from Blaze.

'I'm not completely sure,' he replied. 'But I think I got through to her. She's confused right now. My bond is not as open as it is normally. Don't move. I'll come and join you. If we do it together, I think she'll be all right.'

Elian waited while Tarl moved slowly across the cavern to where Elian was crouching. Together they then crossed the short distance to the remaining egg. Blaze watched their every move intently, gently crooning the whole while. Whether the noise was an expression of her physical pain from the blow to her mouth, or from sorrow at the egg that would not hatch, Elian could not tell.

Elian's ribs and right arm hurt from the impact of the dragon's strike, and the egg was both enormous and heavy. Even with Tarl lending his strength, they struggled to lift it. But no sooner had the egg made contact with the top of the plinth than it seemed to freeze in place. Elian and Tarl stepped back quickly and the same transparent resin-like substance that had formed the outer surface of the Dark Orb and the Orb of Vision oozed from the top of the metal stand. It spread rapidly around the egg until it formed a complete layer.

As with previous orb formations, the clear

substance began to contract. It squeezed the egg, crushing inwards to make its contents smaller and smaller until all that remained was a glowing ball of gold that looked precisely the same size as the previous orbs.

Only when he was sure that the plinth had completed the transformation did Elian reach out to touch it. What would he feel? What qualities would it possess? His hand hovered over the orb for a moment, his arm trembling with anticipation. Taking a deep breath, he grasped it and a shock ran through his body that took his breath away.

'Oh, my word!' he breathed. 'Oh ... my ... word!'

Chapter Twenty-Three
Missing in Action

'*Is that who I suspect it is?*' Elian asked as the familiar throbbing buzz of a flying machine approached through the pre-dawn murk. They had deliberately landed in a quiet area a long way from the lines of fighting. There was no need to risk getting entangled in the fighting again. If it was not for the enormous time saving that moving through this world offered in travelling vast distances across Areth, Elian would not have let Aurora bring them back at all.

'*Yes, Elian,*' Aurora confirmed. '*It's Jack.*'

'*But what's he doing here?*' Elian asked. '*We did what he asked and he led us to the final orb. I thought that was the end of our association with him. How does he even know we're here this time?*'

'*That is puzzling,*' Aurora said, her voice thoughtful. '*It appears Jack has developed a curious sensitivity to* our presence. He is concentrating very hard. He knows we're close and that we dragons can read his surface thoughts. From what I sense of his intentions, he plans to come with us to Areth.'

'But why? I don't understand.'

'It appears Jack feels he can help us get past Segun,' Aurora said thoughtfully. 'He has seen the gateways before. He's determined to fly into this one when I form it.'

'Will his aircraft pass through all right?'

'I don't see any reason why it shouldn't,' she said, with the mental equivalent of a shrug. 'I can't see how he can hope to help us against the night dragons with his one flimsy machine, but he seems very convinced of his worth.'

'Do you think we should stop him?' Elian asked. 'If he crosses to Areth, there's no guarantee that he will ever be able to return to his world. Without wanting to sound pessimistic, it's not going to be easy to run that blockade. If we don't make it, there'll be no way for him to get back.'

'He seems to be aware of the risks, Elian,' she said. 'The Oracle made this man a part of our quest. I think we would be foolish to turn his offer of help aside, even if we cannot see an obvious role for him right now.'

'Kira won't be happy,' Elian noted. 'She doesn't trust him. Not since the incident with the Red Baron.'

'It's not Kira's decision. It's ours. I say we let him come. I've got a strange feeling this is meant to happen.'

'After what happened at the enclave, I'm not about to go against your instincts,' Elian declared. 'I can feel the dawn approaching. We'd better get airborne.'

'You are quite right, Elian. At this rate you'll be developing a dragonsense of your own, soon,' Aurora noted.

'I'm assuming that's a joke,' he replied. Although her voice had sounded serious, he knew her to have a dry sense of humour. She did not reply, leaving him to wonder. Instead, she turned to face across the open field in preparation for take off.

Elian glanced across to where Kira was ready on Fang. She was signalling frantically and pointing towards Jack's approaching flying machine. As he had anticipated, she did not look happy. Giving her a wave and a thumbs-up signal, he gave Aurora the word to launch. Kira would be livid if she found out that his apparent misunderstanding of her signals was deliberate, but he knew that if he played innocent, she would fall for it. She still considered him more naïve than he was, and he had learned there were times when he could use this to his advantage.

Aurora leaped forward, accelerating rapidly until she gained enough speed to take off. The field was not large and the high hedge at the far end was approaching fast when she drove her body from the ground. Elian took a sharp intake of breath as his dragon skimmed the top of the greenery and powered upwards. When he looked back, Fang was safely airborne and following close behind.

Jack's flying machine had entered a gentle turn above them. Elian could just make out his face looking down over the side of the cockpit. The pilot was watching both the dragons and the sky immediately ahead of them.

'Clever,' Elian muttered. 'He's making sure he's in position to make a dive for the vortex in case we try to go through without him.'

'It would not work,' Aurora said. 'I could block him if I wanted, but his determination is admirable. Dawn is upon us. Let's go to the meeting place. Hopefully the others will be there already.'

Elian felt power building inside his dragon as the moment of dawn approached. As she drew it in, he felt a giddying rush of energy surge through the bond that made him feel as if he were swelling inside to the point where he might explode. The outpouring as Aurora formed the gateway was similarly intense. Elian did not need to look ahead to know where the vortex was forming. He could feel it in every fibre of his body.

Fang overtook them, powering ahead to dive into the swirling round disk of grey cloud. As she disappeared, Jack's machine whistled over Elian's head, diving in front of Aurora to follow close on Fang's tail. The wind was screaming through the wires as he pushed his aircraft to its limits to make sure of his goal. Elian caught a brief whiff of hot exhaust fumes in the moment before he and Aurora entered, and then the familiar wrenching twist and the feeling of floating in an eternity of nothingness carried him between worlds.

They emerged to a warm late summer afternoon above the planned rendezvous point. Jack's aircraft wobbled in front of them as he struggled to recover and regain his focus on controlling his machine.

'Is Jack all right, Ra?'

'He is fine, Elian,' she replied. 'The sensation of transfer is always more intense the first time. He is a little shaken, but his thoughts are settling. We should land and give him a chance to recover properly.'

Although her words were positive, Elian could sense Aurora was worried.

'So if Jack's OK, what's wrong?' he asked, getting straight to the point.

'There's no sign of Shadow or Firestorm,' she replied. 'I had hoped they would be here by now, preferably with a lot of back-up.'

'How much time do we have left?'

'Tomorrow night will be the harvest full moon,' she said. 'We need to get the orbs to the Oracle by sundown tomorrow.'

It was a sobering thought. They had a single day to get the orbs to the Oracle. However many dragons Segun had blocking their path it would be too many unless they got help from somewhere.

They descended in a lazy spiral towards a large meadow on the western shore of the lake. They were far enough from the mountains for Elian not to fear any trouble from Segun and his followers.

The dragons landed easily next to the water. Elian and Kira dismounted quickly, both watching with interest as Jack made a more cautious approach and touchdown.

'What's *he* doing here?' Kira asked, her voice carrying a dangerous edge.

'He wanted to come,' Elian replied. 'Ra's dragon-sense was telling her to let him. She thinks his part in the quest is not over yet.'

'But that's ridiculous!' Kira exclaimed.

'That's what I said about waiting while the eggs were hatching,' Elian reminded her. 'Ra's instincts have proved true on more than one occasion. I'm not going to make the mistake of doubting her again.'

Kira crossed her arms and clenched her teeth tightly as the airman's machine trundled to a stop a few yards away. The engine sputtered and fell silent. Jack was out of his straps and out of the cockpit within a matter of seconds.

'What ho!' he called, jogging across to meet them. His eyes were sparkling with excitement. 'This place is amazing – so bright and vibrant! I can't believe I'm actually here – in your world. Thanks for letting me tag along. I confess I panicked for a moment while I was in that strange place between worlds. I thought I'd got stuck in the void. Can't tell you how relieved I was when I arrived.' He craned his neck and scanned the sky again. 'Not as many dragons here as I'd imagined,' he added.

'What did you expect?' Kira retorted, her words clipped. 'A sky full of dragons? We told you dragons were rare here.'

'Yes, you did, Kira,' he agreed. 'And I believed you. It's just . . . another world. The air smells different. Fresher. Cleaner. And it's peaceful. Beautiful!'

Jack kept looking around, his eyes never resting, as if he were trying to soak up every detail.

'Why did you come, Jack?' Elian asked. 'And how did you find us? We were trying not to advertise our presence.'

'I felt you arrive,' he said slowly. 'It was strange.

I knew the instant you came through the vortex into France last night. At first I thought it was my imagination, but inside I knew. Then I had the most vivid dreams about where you had landed and where you were camped. I can't explain why exactly, but I knew I *had* to find you – to come with you. I'm not sure why, but it feels right for me to be here.'

'But you might never be able to go back,' Elian persisted.

'I've spent every day for the past three years wondering if I would survive to see another,' Jack said, smiling sadly. 'The risk of coming here did not seem any worse. So many of my friends are gone.'

'But your family . . .' Kira began.

'Will be told I am missing in action when I don't return from the early sortie,' Jack said, shaking his head. 'Who knows? By the time I do get to return, maybe the war will be over.'

'You may have exchanged one danger for another,' Elian said. 'We need to complete our quest by sundown tomorrow, or the Oracle will die. We don't know yet how many hostile dragons are blocking our way, but our chances of survival are not looking good. We were hoping we would have some friendly dragons waiting here to help us. It looks like our friends didn't make it.'

'Do you want me to go and scout for you?' Jack offered.

'There's no need,' Kira answered abruptly. 'I'll do it with the Orb of Vision.'

A short while later the three sat by the water's edge. Their mood was sober.

'Over three hundred,' Elian whispered again. It was the third time he had said the number aloud. 'How can we get past three hundred dragons?'

He picked up a flat stone and sent it skimming across the water. It skipped five times before sinking below the surface with a final skidding splash. He picked up another.

'A diversion?' Jack asked. 'Something to draw them off?'

'Yes, but what?' Kira snapped. 'We had this argument with Pell last time we considered running the blockade and there were only fifty of them in our way then. There's nothing that we can do to draw and hold the attention of that many. Whatever we do, Segun will just despatch an appropriately sized group to deal with it.'

'We would need an army of dragons to catch his attention!' Elian muttered.

'Like this one, you mean?' Aurora interrupted.

Elian looked around startled. From the look on

Kira's face, she had just had a similar message from Fang.

'What? . . . Where? . . . Good grief!'

'Now that's more like what I was expecting to see!' Jack exclaimed. 'What an amazing sight that is!'

'Amazing is right!' Elian agreed, pumping his fist into the air with joy. 'Yes! They did it! Pell and Nolita did it!'

The sky to the south was full of approaching dragons – all of them blue, except one. Shadow swept down and skimmed across the lake, turning to pass just over their heads and touch down behind them. All three ducked as she flew over them. His dramatic arrival was repeated by Firestorm who, in turn, was followed by another day dragon.

Suddenly there were dragons landing all over the meadow – dozens and dozens of them. Elian tried to count, but kept losing track. He started laughing with happiness. He caught Kira up in a hug and danced around in a circle. He was grinning so hard his face felt as if it might split in two. The day dragons had come. The day dragons had come!

Together, Elian and Kira ran to greet Pell and Nolita and the four riders drew together in an enthusiastic huddle. Kira even hugged Pell, which was something she had never thought she would do. Jack followed Kira and Elian at a more sedate pace, even as Barnabas followed Nolita and Pell from the opposite direction.

'Did you get it?' Pell asked urgently. 'Did you find the egg?'

'Yes, we did,' Elian replied, completely astonished. 'How did you work out it was an egg?'

'Barnabas told me,' Pell confessed. 'He'd solved the entire riddle poem by the time I got to the day dragon enclave.'

'By the time *you* got there?' Kira asked pointedly. 'What about Nolita?'

'It wasn't Pell's fault,' Nolita said softly. 'We were chased by night dragons and split up. We made our way there independently.'

'Were you all right with that?' Kira asked, clearly amazed.

'Not really,' she admitted. 'But Firestorm helped me through it. We survived.'

'Both of your colleagues did very well,' Barnabas said, from just behind Nolita. He placed a hand fondly on the blonde girl's shoulder. 'You can be proud of them. Pell was brave to approach the day dragon enclave alone, and I'm sure you can appreciate how brave Nolita has been. But who is your friend? That is a strange-shaped contraption sitting between Aurora and Longfang. Does that belong to you, young man?'

'Jack Miller's the name, sir,' Jack replied, reaching out his hand. Barnabas instinctively completed the greeting by grasping the proffered hand. 'It's a pleasure to meet you. And yes, the aircraft is mine.'

'Aircraft? It flies?'

'Indeed, sir. It flies nearly as well as your dragons.'

Elian raised a hand to interrupt. 'I hate to break up the introductions, but we need to make some big decisions. Kira scouted Segun's forces earlier and she estimates he has over three hundred night dragons around the Oracle's cave. How many dragons have you brought with you, Barnabas?'

'We have around a hundred and sixty,' he said. 'Enough to draw Segun's attention, but nowhere near enough to take on his forces in any sort of conflict.'

'Wait a moment,' Jack interjected quickly. 'If you can answer some questions, I might be able to teach you tactics that could even the odds a little.'

Chapter Twenty-Four
Confrontation

The arrival of a dozen dusk dragons as the sun went down caused a stir of excitement amongst the day dragonriders. Without warning, a large vortex appeared over the lake and four dusk dragons flying in tight formation had already emerged before the warning cry went up. Everyone stopped to watch as more and more flew out from the swirling disk of cloud. Last to emerge was a single dawn dragon, looking decidedly shaky on his wings as he glided to land at the edge of the lake.

'It's Shimmer!' Elian exclaimed excitedly to Barnabas. 'And his rider, Neema. It looks as if he pushed himself to the limit to bring as many dusk dragons as he could. I know how tired Ra became when she took three of us through the gateways.

Goodness only knows how Shimmer will be feeling after bringing that many through.'

'It was a brave effort,' Barnabas said. 'And a most welcome one. Jack, will having twelve invisible dragons change your tactics?'

Jack's eyes lit up and he gave a wicked grin. 'It won't change what we've discussed so far, Barnabas,' he replied. 'But it's going to give the leader of your enemies a severe headache. Can we gather all the riders together and sit them down? If I talk them through tomorrow's tactics now, they can think about the plan overnight. I then suggest that we have a final briefing in the morning before we go. Any objections, thoughts or questions can be raised then.'

'I can see why the Oracle drew you here,' Barnabas observed, nodding and smiling. 'Will you fly with me tomorrow on Wiseheart?'

'It would be an honour, sir.'

'Very good then. Let's get the men organised.'

'My Lord, one of our ranger patrols reports a large gathering of dragons approaching from the South.'

'Ours?' Segun asked, one eyebrow rising quizzically. He took a bite out of a hunk of bread. 'I thought most had arrived,' he mumbled through his mouthful.

'No, my Lord,' the rider answered, his voice apologetic. 'The patrol says they're day dragons, my Lord – well over a hundred of them.'

Segun stiffened and stopped chewing momentarily as he digested this surprising bit of news. Then his jaws began to move again and he appeared to relax. He had risen before sunrise, anticipating the four questors would try something clever at dawn. When nothing happened at sunup, he found himself almost disappointed.

'I suppose I should've expected this,' he mused aloud. 'Barnabas always was an interfering old fool. He's overstretched himself this time, though. Did the patrol report any other dragons with them?'

'No, my Lord,' the messenger said, his eyes down. 'But that is not to say there aren't any. I get the impression that the patrol did not want to get too close and they came straight back here with all speed.'

'In other words, they ran away!'

'Erm . . . yes, my Lord. It could be seen that way, my Lord.'

Segun's eyes narrowed with anger. He cast the remainder of his bread into the fire and climbed abruptly to his feet. The messenger instinctively tensed, but did not back away. He knew better than to do that. The night dragon leader was well known for his sudden fits of rage, and for punishing messengers who brought news he did not like. But, above all else, he was known to detest cowardice. Those designated for messenger duties often drew lots to determine who would work for which senior rider each day. Unsurprisingly, the short straw always got Segun.

For a moment it seemed that the night dragon leader was poised to order a punishment for the patrol, but to the messenger's surprise, he did not. Instead, Segun shook his head slightly and picked up his black cloak. Flicking it around in a swirl, he settled it over his shoulders.

'No matter!' he announced. 'We'll be ready for them. Tell the men to mount up. Go to the Oracle's cave. Tell those maintaining the vigil at the entrance that they're to stay alert on pain of death. No one must enter the cave until after sundown. No one! Is that clear?'

'Yes, my Lord. Totally clear, my Lord.'

'Good,' Segun said. 'Spread the word as you go. Today will be a glorious day for the night dragon enclave. Not only will we end the Oracle's domination of dragonkind – we'll cripple the day dragon enclave. Things could not be better.'

Delighted to be released unscathed, the messenger raced away to do Lord Segun's bidding. The tall dragonrider watched him go, a sneer of contempt on his lips.

'Well, Widewing,' he called to his dragon. 'Are you ready to battle your day dragon brothers and sisters?'

'It has been a long time coming, Segun,' she replied eagerly. 'Let's go! Climb onto my back so we can usher in the new Age of Dragons together.'

'Here they come,' Barnabas called over his shoulder to Jack. 'Dead ahead. Ugly lot, aren't they?'

To Jack, looking past Barnabas and the rising and falling head of Wiseheart, the huge black smudge rising into the sky from the mountains looked rather like an enormous flock of birds, or a swarm of insects. The only difference was the scale involved.

He looked around at the orderly flights of dragons he had arranged. In contrast to the swarm of night dragons launching ahead, the day dragons were flying in tight 'V' formations of three. Each three-dragon 'V' was in turn grouped into a larger formation, made up of five sets of three dragons. Ten of these groups made up the bulk of the force. The remaining dragons were flying above them, to act as spotters, each tasked with coordinating a particular group. Barnabas and Jack were operating in this role.

The dusk dragons had already camouflaged and

were flying wide on the flanks, working in five pairs. The final pair of dusk dragons had been tasked with accompanying Fang, who was carrying both Kira and Elian on their bid to deliver the last two orbs. Aurora and Shimmer were flying above and behind the rest of the dragons. Nolita and Firestorm, together with another day dragon and her rider, were flying alongside the two dawn dragons to protect them against any night dragons that broke through the main lines.

'Good Lord!' Jack said. 'This is going to be the mother of all dogfights!'

'It certainly looks that way,' Barnabas replied. 'Who knows, we may even end up trading blows with Segun himself shortly.'

'Oh, I hope so,' Jack whispered, cupping the butt of the Lee-Enfield rifle strapped across his back with his right hand. 'I do hope so.'

'They look impressive, don't they,' Elian said, glancing across and up at the orderly formations of day dragons.

'Let's just hope that Jack's tactics work,' Kira replied, not looking around. 'Personally, I think he's out of his depth when it comes to dragons.'

'He's spent a lot more time in the air than we have, Kira. Don't let your one bad experience with him blind you to his abilities. He's been fighting in the air for years. He's a survivor. None of the dragonriders has his experience. Look! The night dragons are rising to meet the day dragons. It's working just the way he said it would.'

'Segun won't be foolish enough to commit all his forces,' she pointed out. 'We'll still be outnumbered when we reach the Oracle's cave.'

'I'm sure you're right,' Elian agreed, not wanting to annoy her further. She was focused on getting her orb to the Oracle. He wished he could be so single-minded. Being separated from Aurora did not help, but the day dragons needed her abilities now. He scanned the sky looking for her. She was too far away for him to reach her with mental communication, but he could just about see his dawn dragon above and behind the main body of day dragons.

'*Fly safe, Ra,*' he thought, attempting to cast the message across the gulf of space between them. '*Don't do anything foolish.*'

'Do you think Pell made it without being detected?' Elian asked Kira the question that had been on his mind since Pell set off alone earlier.

'I hope so,' Kira replied, shaking her head. 'He's brave, that's for sure; let's just hope his plan works.'

'It does makes sense,' Elian insisted. 'Jack said that

an enemy is like a snake – cut off the head and the body will die – let's hope Pell can get to the night dragons' head.'

As Elian concentrated again on the scenery around them he saw the mountains were looming close, and their chosen pass towards the Oracle's cave was ahead. It would take them a long way around, but they were determined not to draw attention to their approach.

'There's a night dragon on the right side of the pass,' Kira warned. 'We're going to climb and go above him. He's less likely to notice us that way.'

'And so it begins,' Elian muttered.

'There they go,' Pell observed, peering out from his hiding place under the thick pine trees. *'Just as Barnabas predicted. Let's wait until the majority has passed before we launch. Can you make out Widewing amongst all those dragons?'*

'No, not yet,' Shadow replied. ' *But I imagine he will be towards the back of the main force. Segun likes to direct from a point of safety, rather than lead from the front.'*

'Barnabas read Segun perfectly,' Pell added. *'I thought the old day dragonrider might attempt a parley, but he was sure Segun wouldn't want to talk. He predicted Segun would launch straight into a full-scale* attack. He was right. There's no way they'll avoid a fight with that many dragons in the air. Do you think Jack's other-world tactics will work?'

'They are novel enough to give the day dragons the advantage of surprise early on,' she said. 'But the superior numbers, size and sheer momentum of the night dragons will break them. The day dragons have ever been brave, but their attempted show of strength is about as clever as a human taunting a lion with a bleeding haunch of meat. The end result will be ugly.'

Pell was not quite so sure. Having seen the result of Firestorm's unorthodox tactics against the night dragon in the mountains ten days earlier, he could see how Jack's tactics would prove devastating if executed well. If the day dragons could inflict enough damage in the first few moments, the momentum could change very quickly.

He continued to watch as black wings darkened the sky above. Finally he judged the time to be right.

'Ready?' he asked.

'It is time,' Shadow confirmed.

Pell ran to her side, leaped up into the saddle and hooked his arms and legs through his newly fitted fighting straps. No sooner was he settled than Shadow started to move towards open ground. Even before she had fully emerged from the cover of the green canopy, she began to accelerate into a run. They burst from the trees at speed and within a few more paces, Shadow had extended her wings and they were in the air and climbing.

As Pell had expected, none of the night dragons, or their riders, noticed them climbing up to join the throng from beneath. Everyone was focused on the incoming lines of day dragons. They powered up amongst the other night dragons without attracting any attention to themselves, but Pell knew all too well that this was the easy part of what they were trying to do. Somehow they needed to single out Widewing quickly, preferably before the two sides clashed, though that appeared unlikely. At current closure, Pell could see that, at best, they had a couple of minutes before the fighting began. When that happened, he and Shadow would be in danger from both sides.

Once amongst the huge flight of night dragons, the noise was incredible. The *whooshing* sound of a dragon's wings had been special to Pell from the first moment he had experienced it. Flying in formation with Fang, Aurora and Firestorm had added a different dimension to the sound, but this . . . this was like a raging ocean in a hurricane, though even that comparison did not do the sound justice. It was awe-inspiring to think that he was flying into the midst of what was possibly the most dangerous force on Areth.

Aside from the sound, the air was choppy with downdrafts from the wings of dragons above them, and turbulence from the passage of dragons ahead. They bumped and jostled through layer after layer of dragon wings, weaving and dodging to work their way ever higher.

'*I see them,*' Shadow announced suddenly. '*Widewing is ahead and slightly to the right.*'

'*Great,*' Pell replied. '*Let's get into position to strike.*'

Chapter Twenty-Five
Dragon Battle

'They're still climbing!' Segun muttered, his thoughts spilling from his lips. 'What's Barnabas up to? He must know the day dragons can't outclimb us.'

The night dragon leader looked at the mass of night dragons ahead, below and all around him. His superior force was moments away from combat and he could feel his heart rate accelerating.

'*Are you ready?*' he asked his dragon.

'*I am,*' she replied confidently.

Suddenly, a brief flash of light from above and behind the approaching day dragons drew her attention. Segun was looking to his right and felt, rather than saw, it. What had caused it? He narrowed his eyes against the wind and tried to force his focus out beyond the approaching formations to see if he could pick out the source. Without warning, he found his eyes dazzled by the most intense, burning light he had ever known.

To the night dragons and their riders it was as if twin suns had suddenly exploded into being in the sky ahead. The glare was incredible and the timing, crippling. Just before the combat began, two flares in the sky blinded almost every night dragon and rider. Segun and Widewing were no exception. The flare burned fierce and bright for several heartbeats. Even through his eyelids, Segun could not totally block out the light. Then one, followed shortly afterwards by the other, died away to nothing.

Hands over his eyes and trying to rub away the flash spots that were dominating his vision, he began to swear and curse. Even as he did so, a dragon crashed into them and Segun suddenly found it was all he could do to stay in the saddle as Widewing was dragged into steep spiral dive. The world spun and the wind roared in Segun's ears as they accelerated. A sharp pain erupted in his lower back and he felt further brushing collisions as they fell together through the layers of dragons beneath.

It took several moments to realise that the pain in his lower back was actually Widewing's. His back was uninjured, but he was experiencing her pain through their bond. They were being attacked! For a brief moment he thought it must have been a dusk dragon that had used its camouflage to get in close, but then he caught sight of the dragon grappling them in his peripheral vision. It was most certainly another night dragon.

'*Widewing!*'

'*It is Shadow and Pell,*' she replied, her voice strained and thick with anger. '*Hold on tight. I'm going to dislodge them.*'

Had Widewing not warned him, Segun would have almost certainly been flung from the saddle. He instinctively grabbed for the straps and a couple of quick twists of his wrists secured him in place just in time for Widewing's desperate manoeuvre. With stomach-wrenching abruptness, she lurched into a spiral dive.

Shadow had sunk her talons deep into Widewing's tail and was biting at her flank and lower back. Folding her great wings in tight to her body, Segun's dragon made herself a dead weight. At the same time, she twisted to strike at Shadow's exposed neck. The twisting motion whilst they were already spinning threw Segun to the left so hard that for a moment he felt as if his arms were being pulled from their sockets.

Widewing's tactic succeeded in dislodging Shadow, but the pain remained. Pell's dragon had penetrated Widewing's armour with her talons and teeth in several places. Segun had never known his dragon to feel such anger. They continued to freefall for two or three heartbeats before Widewing extended her wings again and swooped up to meet her opponent.

Shadow was ready for the move and the two dragons met with talons raised and mouths open, poised to strike. For a moment they both hung there, wings beating hard to maintain the hover, talons raking and great teeth biting, each trying to get a hold that would give an advantage. The moment passed. Neither dragon could hover for more than a handful of wing beats.

Pell's dragon was first to break away, diving suddenly underneath Widewing and racing off in the direction of the mountains.

'The Oracle's cave!' Segun called out aloud. 'After him! He must not be allowed to reach it!'

'*Now!*'

Shadow broke from her hover and dived underneath Widewing. Pell ducked and felt the talons of Segun's dragon whistle through the air just above him. They were clear. He looked over his shoulder. Sure enough, Segun had already wheeled Widewing around and they were in pursuit, just as he had hoped.

Overhead the night dragons were having a terrible time. Many were spiralling out of control towards the ground with great smoking holes in their wings. Jack's tactics appeared to be working exactly as he had said they would. There were a few night dragons grappling with day dragons, but in the main, the day dragons were using their fire to make effective hit-and-run attacks. By aiming their attacks at the one weak point of the night dragons – their wings – the day dragons were causing havoc, sending them spiralling to the ground in large numbers.

It had been Elian's idea to use Aurora and Shimmer to blind the night dragons and their riders. Jack had talked about keeping the sun at their back to make it difficult for the night dragons to see, but Elian had pointed out that this would not be possible when approaching from the west, as they were. The surprise factor of having two sources of incredibly bright light in the form of the dawn dragons was even better.

In a matter of two or three heartbeats, the two dawn dragons had rendered almost the entire host of night dragons and their riders temporarily blind and totally helpless. They could not see where the attacks were coming from. Many collided with each other in the chaotic scramble to avoid the day dragon fire that was suddenly raining down on them.

The pairs of dusk dragons attacked the flanks, causing more mayhem. The outer edges of the host were forced in on one another by their random strikes. As the night dragons bunched ever more closely together, so they became easier targets for the fire of the day dragons, and more accidental collisions occurred.

Pell's mission was to keep Segun and Widewing out of the fight. He was determined to keep them too angry to notice what was going on. Without their leader to rally them, the night dragon attack was less likely to regain cohesion.

'We've got their attention, Shadow!' Pell noted, as the gigantic black dragon pounded after them with fearsome power. *'Now would be a good time to fly like the wind. And don't look back!'*

'We're getting close,' Kira said over her shoulder to Elian as they rode together on her dragon. 'Fang says he can sense at least half a dozen night dragons still in position around the Oracle's cave and possibly more inside. We're going to hold here while our wing riders try to draw them away from the cave. Fang won't stand much chance if it comes down to a straight fight. The night dragons are too big and powerful.'

'Thanks for the update,' Elian replied, projecting his voice into the wind.

Even with the day dragons distracting most of the night dragon enclave, the odds of getting past the remaining dragons were not great, but Elian knew Kira and Fang would do everything in their power to reach the Oracle.

Fang entered a turn over the meadow where the dragonhunters had attacked them when they had been preparing to first set out on the quest. It was hard to believe that only six weeks had passed since Kasau had brought his men to kill their dragons. It felt like a lifetime ago that Shadow had eaten the strange dragonhunter. They had travelled so far and seen so much since that fateful evening.

Elian waited patiently. As he did so, he became aware of the orb in his backpack radiating its amazing power through the leather. He knew it would not be easy to drop it into the dark well of the Oracle. If he kept it, the orb could be used to help so many, but if he did not relinquish it then the Oracle would die.

His first contact with the final orb had been an experience he would never forget. The shock of its power had filled his body: healing, revitalising, re-energising. It was akin to the healing flame of

Firestorm, but infinitely more potent. His eyesight had returned to normal in an instant, as had his hearing. The scar on his leg from the blast wound had disappeared and the muscles had regenerated. There was no longer so much as a hint of pain. His body had not reversed in age, but he felt fresh and clean, as if he had been made anew.

Elian had taken the orb out of the hatching chamber and through the caves to the space where Aurora and Longfang were waiting. Kira had followed, not sure what he was going to do, but sensing something momentous was about to happen. Elian had lifted the orb and touched it against the side of Longfang's head. In the brief instant between one heartbeat and the next, the orb's power had surged through Longfang and miraculously regenerated his missing eye. Kira had cried out with shock and delight, before running to embrace first Elian and then her dragon, tears streaming down her cheeks.

It was easy to imagine how the power of this amazing orb could spark the rebirth of the Oracle. The choice between using the orb's power to help others, and sacrificing it to see the Oracle reborn, had been the hardest decision of his life. He was still not sure he had made the right decision, but his personal choice might become irrelevant if they could not get past Segun's henchmen.

'Mikhael and his dusk dragon, Spirit, have successfully drawn away all the dragons guarding the outside of the cave,' Kira called over her shoulder. 'Flicker and Ines are going to draw out as many as they can from inside.'

'That's great!' Elian enthused. 'Perhaps we should follow and lurk above the entrance ready to make a dive for it when Flicker and Ines have done what they can.'

'Fang thinks that's a good idea,' Kira told him. 'Hang on. We'll follow above and behind them.'

Rolling out towards the familiar entrance to the narrow valley, Fang began to climb. It was hard to make out Ines. She had deliberately worn grey leather to try to blend in with the rock and her choice of colour had been a good one. Her dragon, Flicker, was as invisible as Fang.

Currents of air bumped and buffeted them as they entered the narrow valley with its steep rocky sides. The wind was not strong, but it was brisk enough to cause significant turbulence. The valley opened out slightly near the Oracle's cave, Elian remembered. This was good, for there was not enough space here in the valley for even a relatively small dragon like Fang to circle safely. They bumped and bounced in the choppy air, but the lumps and bumps of the turbulence no longer sparked any fear in Elian. He was more concerned about what awaited them inside the Oracle's cave.

They wended their way along the valley until the dark hole in the cliff face became visible. Using the widest point, Fang entered a steep turn to hold his position while below and ahead of them, Flicker swept down to land on the ridge at the cave mouth. Elian could just make out Ines lying flat on her dragon's back as he entered the cave. They were not inside long before Elian and Kira heard the sound of distant roars.

Elian did not see Flicker emerge, but there was no missing the two night dragons that raced out and launched from the ledge with mighty leaps. They were quick to turn away from the area in which Fang was circling. Flicker was deliberately leading them away.

Fang wheeled around until he was flying straight towards the Oracle's cave and entered a shallow dive. The wind rush increased as they accelerated towards the black hole in the rock, and as the grey rock wall began to swell towards them with ever increasing speed, so Elian began to wonder if Fang had quite got used to having both his eyes again. It felt as if they were going too fast to land safely, and Elian's heart rate accelerated. Fang dipped the rear end of his body like a giant airbrake. Altering his wing angle sharply, Fang gave three mighty back wing flaps to complete the braking action and landed with barely a bump on the ledge outside the cave.

'Quick! Dismount,' Kira ordered. 'Fang will lead the way in. If there are any more dragons inside, he will do his best to deal with them while we deliver the orbs.'

Elian did not hesitate. He was sliding down Fang's side and up against the rock wall of the cliff face before she had finished her sentence. Kira landed lightly beside him, and Elian heard the faint clicking of Fang's talons on the rock as he moved away to enter the cave.

Kira looked him in the eye, her expression holding a challenge as she drew two knives from her belt – one for each hand. 'Best to be prepared for the worst,' she said.

Elian reached back over his left shoulder and drew his dragonbone sword with his right hand. 'Right,' he agreed softly. 'Do you want me to lead?'

'No, I'll go first,' she whispered. 'No offence, but I move more quietly than you do. Be ready to back me up if I run into trouble.'

'I'll be right behind you,' he assured her. 'Let's go.'

Chapter Twenty-Six
Sacrifice

'They're gaining on us, Shadow!' Pell yelled urgently. 'They're right on your tail!'

'*Hang on!*' she replied. '*I can't go any faster. We're going to have to turn and fight.*'

The first line of mountains loomed not far ahead. They had run a long way from the main battle in a remarkably short space of time. A turn would become inevitable shortly anyway, as there were no convenient passes ahead. Great rock-strewn slopes with sparse patches of green towered in front and the textured blend of forest greens spread immediately below them.

Shadow had avoided turns to prevent Widewing from cutting corners to catch them. Now she was left with no choice. The mountainside was rearing up in front of them. They would not be able to outclimb

Segun's dragon, and Widewing was so close she was likely to catch them the moment they entered the turn. As a result, Shadow was hesitant, not wanting to commit one way or the other.

Dipping her wing in a quick feint to the left, she tried to trick Widewing into committing to the turn, but before her direction had begun to change, she reversed rapidly to the right. To Pell's horror, Segun's dragon was not fooled. The enormous night dragon struck at Shadow's rear right quarter, its teeth scoring a painful tearing wound through her scales. Before Shadow could recover, Widewing grabbed her tail, slewing her around so abruptly that even with his strong grip on the straps, Pell had to fight to stay in the saddle.

He was terrified. An explosion of fury and pain suddenly blocked his link to Shadow's mind. Shadow rolled under him, twisting to strike back at Widewing and the two dragons began to tumble in a grappling, spinning spiral once again. Occasional flapping by either dragon slowed their fall in lurches, but the decelerations were never more than momentary and their momentum was still carrying them towards the mountainside.

The rocky mountain slope was so close that, as they fell, it felt to Pell as if the entire mountain was sliding sideways towards him. He began to yell warnings, but Shadow was so focused on her grappling fight with Widewing that she did not appear to hear him.

In the instant before they crashed into the slope, two words blasted through the bond.

'JUMP CLEAR!'

Pell barely had a moment to prepare. Unwinding his wrists from the straps and kicking his feet clear of the stirrups, he freed himself from everything that could prevent him from leaving her back. There was the briefest moment of anticipation before . . . crunch! The impact was terrible. Both Shadow and Widewing hit the steep scree slope together and immediately began to roll, still biting and clawing in the midst of an avalanche of loose shingle and rock.

In the last few heartbeats before they struck the ground, Pell flattened himself against his dragon to spread the expected impact across as much of his body as he could. At the instant they struck the mountainside, Shadow gave a sudden arch of her back that catapulted him into the air. Already winded by the first bone-crunching collision, Pell's second landing felt even harder. He smacked down onto the scree slope, and immediately beginning to slide. For a brief moment he thought he was slipping across the surface of the steep field of small stones and rubble, but to his horror he realised something much worse was happening.

The rolling fight of the two dragons had not stopped on impact. Already accelerating away down the mountainside in a cloud of dust and falling stones, the two dragons remained locked in combat. But their passage had triggered a secondary effect. The entire upper portion of the slope had been set in motion. Although most of the stones that were moving were barely more than gravel, the sheer mass involved meant bigger stones and rocks were being swept into the fall as well. Pell was riding the wave of an avalanche.

Initially he slid sideways, but managed to turn until he was riding feet first down the slope. He dug his heels in hard with the intention of slowing his progress, but it made no difference. The entire surface layer was in motion. All he could do was ride it out and shield his head as best he could from anything travelling faster than he was.

A rush of adrenalin masked any pain he might have felt from cuts and abrasions on his back, bottom and legs. His heart thumped and his breath came in ragged gasps as he tumbled down the mountainside in a storm of stone. Several larger boulders bounded past on either side of him and one bounced directly over his body, only to continue gambolling down the slope, shedding smaller fragments of rock with almost every impact.

Miraculously, as the avalanche slowed, he remained relatively unharmed. He was scratched and bruised all over, but nothing worse. Once he had come to a stop, it appeared he might be buried alive under a sea of the smaller stones. They continued to trickle down around him for some time, along with occasional bigger ones. Wriggling and squirming to stay on top of them, he finally found enough purchase to stand.

Shadow and Widewing were no longer in sight, but he could hear their roars and see the treetops thrashing around where they were still fighting. The two dragons had smashed into the treeline and been swallowed by the forest some distance below. The whereabouts of Segun had not crossed his mind until a sudden rattling crunch to his right caught his attention.

The leader of the night dragons looked terrible and terrifying. He was bleeding from numerous cuts across his face and hands. One of his eyes was already blackening and swelling, and he was limping. His shoulders were hunched with murderous intent and his eyes burned with a fury unlike anything Pell had ever seen before. If the look in Segun's eyes and the set of his shoulders were not enough, the blade clenched in his right hand made his intentions more than clear.

Pell backed away slowly, feeling for his belt knife. His hand found an empty sheath. His blade was gone, no doubt buried somewhere under the rubble. He was unarmed. Quick as a striking viper, he stooped and picked up a stone that fitted comfortably into his palm. Segun did not so much as pause. He kept coming forward, his lips curling into an expression that was half smile, half snarl. His knife hand came up into the classic ready stance as he steadily closed the gap between them.

It was hard enough to walk across the steep field of loose stones, but to do it backwards proved impossible. Despite his attempted care, Pell lost his balance and toppled back. Segun immediately surged forward. Pell's instinct was to throw the stone. Segun twisted his upper body to the left allowing the stone to whistle past. By doing so his right arm was thrust forward leaving him momentarily in a weaker stance. Even as he was falling, Pell saw the opening and swept his right foot around in a crescent kick that caught Segun's knife hand hard. The blade flew from his grasp, spinning away down the mountainside until it skittered to a clinking halt amongst the scree some distance away.

Segun hesitated, his head automatically turning to

follow the flight of the blade. The brief distraction gave Pell just enough time to leap back to his feet and adopt a fighting stance. Segun did not look impressed. The older man was taller and heavier, with a longer reach. He adopted his own fighting stance and Pell could see from the easy way Segun set his balance that the man was no stranger to unarmed combat.

'What's the matter, boy?' Segun taunted. 'Afraid of an old man?'

'Not afraid, Segun,' Pell replied, keeping his focus on the centre of his opponent's chest. 'Not of you.'

'Well, you should be!'

Segun's attack was fast and powerful. His hands were incredibly quick, but Pell had already identified the man's weak spot. He was clearly favouring his left leg. Segun was doing a good job of masking his limp, but it was an obvious target for Pell to focus on and he was quick to exploit it. The older man was attacking purely with his hands. He was quick, but so was Pell. With a rapid sequence of blocks and deflections, Pell prevented Segun from landing any heavy blows before, dropping underneath an inward knife-hand strike, he swept a low turning kick to Segun's right knee.

The kick did not connect hard, but it must have

hit the right spot because the older dragonrider folded over the injured knee and lost his balance. Toppling down the slope, he slid some distance before he came to a stop. Taking care not to trigger a further slide, Pell descended the slope step by cautious step towards the leader of the night dragon enclave. Segun remained face down and still as Pell approached, but his ruse was transparent. Pell stopped well short, determined not to play into the older man's trap.

Seeing that Pell was not going to be fooled, Segun struggled to his feet, making a deliberate play of the weakness in his right leg. Again, Pell could see what was happening, so he held back, keeping his distance and the advantage of higher ground. When Segun did settle his balance, Pell realised his judgement had been better than he realised. There, in Segun's right hand, was the deadly knife again. If he had moved in close, Pell had no doubt that he would have felt the sting of the nasty-looking blade.

'Come on then, boy,' Segun said, his voice still taunting. 'Come and try that move again.'

A movement in the sky behind the leader of the night dragons caused Pell to glance up.

'NO!' he called out, as he realised what was about to happen. 'No, Jack! Don't!'

But it was too late. Wiseheart had broken away

from the main fight and glided down on silent wings towards them. A sudden 'CRACK' reverberated across the mountainside and a blossom of red exploded from Segun's chest. He looked down at the gaping hole and his jaw dropped in a stunned look of amazement that Pell would never forget. With painful slowness, the leader of the night dragon enclave dropped his knife and lifted his hands to his chest. With his face still set with a look of shock rather than pain, he sank to his knees and then toppled forward – dead.

Pell looked up at the two figures on the back of the day dragon as it turned away and a flash of hatred raced through him. What right had Jack to take Segun's life? Pell had determined to exact revenge after the treatment he had received at Segun's hands – not to have some otherworlder come along and do it for him. He felt cheated.

Two screeches sounded from amongst the trees below: one of deep loss and the other of triumph. Pell could feel the latter belonged to Shadow. She had overcome Widewing. It was over.

Elian followed as close behind Kira as he could without compromising her silent advance. Within a few moments she had moved inside the entrance to the Oracle's cave and beyond the pool of natural light from the mouth. It did not extend far inside, but he could see the orange flicker of torches ahead. Kira did not dawdle. She moved forward at speed, but her footfalls were so light that they made no noise. It was hard to keep up but he concentrated on moving as quickly and quietly as he could.

Kira slowed as she approached the alcoves of the guardians, but no one stepped out to challenge them. She paused to look inside the recess on the right, but moved swiftly onwards. Elian copied her quick scan as he passed it, but there was no sign of anything living.

A loud dragon's roar in the tunnel ahead set Elian's pulse racing. Kira accelerated, flitting along the last short section of tunnel to where the great ramp zigzagged down to the chamber floor where the black hole of the Oracle's well awaited them. She stopped at the entrance to the chamber and dropped to one knee as she surveyed the cavern.

Shrugging her pack from her shoulders, she put down her knives and delved deep into the top of the pack. Elian reached her just as she pulled out the Orb of Vision.

'Quick!' she whispered. 'Fang's drawn the two remaining night dragons into the back caves. He won't be able to keep them guessing for long. Drop your pack. We'll just carry the orbs and our weapons. Look. The two riders are blocking the way to the Oracle's well. We're going to have to get past them and we're not going to do that without a fight.'

Elian was surprised to see how dark the chamber was. It was definitely not his imagination this time. The light from the torches on the walls was not as bright as it had been on previous occasions. He swung his pack from his back and quickly rummaged through it until he found the globe wrapped in cloth. Unwrapping it, he held it carefully in his left hand and recovered his sword with his right.

'We won't get down there without being seen,' Kira whispered. 'But I don't want to give them time to think about what they're doing. We're going to do this, and do it fast. Get close to the Oracle's well and lob your orb in. OK? Let's hope the Oracle recovers quickly. We could do with some help if we're going to get out of here alive.'

'I'm ready,' he replied. 'Good luck.'

Kira did not hesitate, she was up and running down the ramp almost before Elian had finished speaking. Although she was running silently, the time for stealth was past. Elian sprinted after her. A shout of alarm sounded from the chamber floor. The night dragonriders had seen them.

Elian caught up with Kira just as they reached the

bottom of the ramp. The two riders, both big men, were waiting for them with weapons drawn. Kira tried to dodge to the left, but the rider on that side was expecting the move. He swung his sword at her in a flat, slicing stroke at waist height. She evaded, twisting her body and chopping down on his blade with her hunting knife. The clang of metal on metal rang loud, echoing around the chamber.

Elian's heart felt as if it had climbed through his chest to his throat. These riders meant business. He did not try to dodge around the second rider. Instead, he adopted the fighting stance that Kira had taught him and edged forwards, poised for the first exchange of blows. A sudden noise on the ramp above and behind him proved a momentary distraction he could not ignore. He glanced back. The other night dragons were returning! They were already out of the tunnel and in the chamber, racing down the upper ramp.

Although Elian's distraction had only been momentary, his opponent took full advantage of the opening he had gifted him. Even as Elian's attention switched back to his opponent an instant later, he realised his inattention would prove fatal. Everything seemed to slow down. He could not block the man's lunge. The sword was driving towards his chest with deadly power. He tried to deflect it, but he was far too slow. The pain exploded as the blade drove between his ribs, straight through his heart and out of his back.

He wanted to cry out, but he could not make a sound. In the distance he heard Kira shout his name, but it sounded as if it were coming from many miles away. A rushing sound of blood roared in his ears. The look on his killer's face was of satisfaction as the blade rammed home. The man paused for a moment, looking Elian in the eyes and then he wrenched the blade free again.

Elian fell to his knees. The pain had already gone. Golden fire burned brightly before his eyes and consumed him. It filled his vision. It filled his mind. It filled his body.

'So this is what it's like to die!' he thought. *'It's not so bad.'*

But then the golden fire began to disperse. Suddenly, he could see the expression on his killer's face again. This time it did not appear satisfied. The man looked awestruck and afraid. There was no pain. In fact, Elian felt perfectly healthy. He climbed back to his feet and raised his blade again. A strange tingling was still racing around his body. The feeling was particularly intense in the area of his wound and in his left hand, but he felt no pain. He felt perfectly fit and whole.

318

'The orb!' he breathed. Realising the man had heard his words, he decided to amplify them. 'I hold the Orb of Rebirth. I cannot be killed.'

He stepped forward, clutching the orb in front of him. The man stepped back, terrified. Elian moved forward again, this time swinging his sword in a testing stroke. The man blocked, but his blade was made of ordinary metal. Elian's dragonbone blade sliced straight through it, cutting the blade almost precisely in half. That was more than enough for the night dragonrider. He fled. Elian turned his attention to the man's companion, who was quick to follow his friend's lead.

'Come on, Elian!' Kira urged. 'There's no time to waste!'

The incoming night dragons were coming down the ramp towards them at full speed. Kira sheathed her belt knife and grabbed Elian's wrist, dragging him towards the Oracle's well. It was no more than twenty paces away. Still struggling to recover his composure, Elian stumbled alongside her until they reached the walled abyss.

Without pause, Kira dropped the Orb of Vision into the darkness and turned to face the approaching night dragons.

'Drop it in! DROP IT IN!' she cried anxiously.

But Elian could not do it. His fingers were clasped

tightly around the orb. He knew this was what the past six weeks of pain and trauma had been all about, but he could not bring himself to let go of it. The orb had given him life when he should have died. How could he let such a thing go?

'Drop it, Elian. You must.'

'Aurora!'

Another challenging roar sounded in the Oracle's cavern. Kira gasped and Elian looked around, his left hand still outstretched over the wall. More dragons were coming down the ramp. The three approaching night dragons stopped and turned. Firestorm was at the top of the ramp. Behind her were Aurora and Shimmer. Elian could see Nolita clinging to Firestorm's back as he charged towards the surprised night dragons. More dragons were appearing – all day dragons.

'Let the orb go, Elian,' Aurora told him again. 'You cannot keep it.'

'But it—'

'I know,' Aurora said, cutting him short. And he knew from her voice that she understood perfectly. 'Do it. Now! You must.'

'Listen to your dragon, Elian!' Kira urged. 'Drop it! Quickly!'

A blast of fire roared from Firestorm's gaping mouth and the three night dragons screeched defiantly. Two more screeches sounded from Elian's right. The other two night dragons were returning. Aurora was right. If he was going to restore the Oracle, it had to be now.

He turned and looked longingly at the glowing wonder in his hand. It was beautiful, but the Oracle needed it. Dragonkind needed it. He relaxed his fingers and tipped his hand slowly, allowing the orb to roll from his fingers. His eyes followed it for a heartbeat . . . two . . . three. Then it was gone.

It was done.

All the strength drained from his legs and Elian sagged against the low wall. He looked around. Firestorm was loosing his fiercest fire in great jets at the night dragons. They had backed away, but showed no signs of running.

'Come on, Oracle!' Kira muttered, looking down into the great well. 'Where are you? We need you!'

The same question was at the forefront of Elian's mind.

'It's not quite over,' Aurora said softly, her voice almost apologetic in his mind. 'There is one last sacrifice required.'

'Sacrifice?' Elian asked, confused. 'What sacrifice?'

'"Beyond time's bright arrow, life-saving breath,"' she quoted. '"Love's life force giving, slays final death . . ."' The Oracle reiterated the need in the final verse with:

321

"Gifted for ever: life's sacrifice." The Oracle is already dead. It knew it would die before we brought the four orbs. The only way it can be reborn is for someone to give up his life force. Someone must die.'

'Die!' Elian exclaimed aloud. 'But the dragonet in the egg died. That was the sacrifice, wasn't it?'

'No, Elian,' Aurora said softly. 'The Oracle chose carefully for this quest. We questors are the ones who have to make the sacrifices. For Nolita it was blood. For Pell it was to be his dragon's heart . . .'

'Wait a minute!' Elian interrupted. 'You mean me, don't you?'

'Not just you, Elian,' she told him. 'We must both make the ultimate sacrifice. It is our life purpose to see the Oracle rebirthed. Dragonkind will be saved through our gift. There is no need to fear. It is a glorious life purpose, Elian. We will live for ever in the memories of dragons and riders everywhere.'

'But I don't want to die!' he told her through the bond. 'Becoming your rider was supposed to be an adventure.'

'And it has been,' she said gently. 'It has just been a little shorter than we expected. We have no choice, Elian. "Love's life force giving, slays final death . . .". Without the final sacrifice, the Oracle cannot be reborn. There does not need to be any pain. Remember your fall from the Devil's Finger? You were ready to die that day. This is

322

the fall you were destined to make. Don't be afraid. We'll dive into the Oracle's well together. Wait there. These night dragons cannot stop us now. We are about to begin our final adventure.'

Elian looked over the low wall at the yawning black chasm below and his stomach churned with fear. Tears welled in his eyes. He did not want to end his life this way. Memories of happy days with his parents came thick and fast in a jumble of images. He remembered playing with friends in his village, his father telling him bedtime stories and his mother serving up special food for family occasions. The roaring fight in the chamber faded as his dreams filled his mind, but could not be totally banished by his reverie.

Suddenly new images intruded – plumes of smoke, ragged lines of tired men marching with their weapons held ready, night dragons swooping down to attack his village, his parents' cottage in ruins.

'Stop it!' he cried out, hitting his temple with the heel of his right hand. 'That's not real.'

'No, but it will be if the Oracle is not reborn.'

'You don't know that for sure, Ra.'

'I know enough of Segun and his followers to assure you life without the Oracle will not be pleasant.'

'What is it, Elian?' Kira asked. 'What's wrong?'

'Ra thinks we're the final sacrifice – me and her,'

323

he croaked. 'She wants us to dive into the Oracle's well together.'

'But that's crazy!' she gasped. 'Surely the Oracle didn't mean that to happen.'

'I've had bad feelings about the quest for a while now,' Elian said, his tears running freely down his cheeks. 'There was something about the way the Oracle acted when Pell and I were here last time that left me feeling very uneasy. I don't want to die, but Ra is sure this is the right thing to do.'

'You can't, Elian! It's wrong.'

'I don't think I have a choice. Everything we've been through has brought us to this moment. If I don't go through with it, Segun will win and the world will suffer.'

'There's always a choice,' Kira insisted. 'There must be another way.'

'*FIRESTORM! NO!*'

Aurora's cry demanded attention. She was locked in combat with a night dragon, trying to force her way through to Elian, but it was Firestorm who had made the breakthrough. The battling blue dragon forced its way through the cluster of night dragons and ran straight towards him. He froze. What did Nolita and Fire think they were doing? Separating themselves from the rest of the dragons was a sure way to become vulnerable. Then Elian realised.

Firestorm was not battling towards *him*. He was heading for the Oracle's well.

A night dragon lunged at Firestorm from the right as he broke through their ranks, but Firestorm ignored the wound it dealt to his hindquarter.

'Get down Elian!' Kira yelled, hitting him with a shoulder charge that flattened him just in time for Firestorm to leap over them and disappear into the black nothingness of the Oracle's well. As the blue dragon passed over them, Elian caught a glimpse of Nolita clinging to his back. She was smiling. Smiling!

'NOLITA!' he cried, scrabbling to get out from under Kira. He looked over the wall, but they were gone – swallowed by the darkness. 'NOLITA!'

Kira's hand grasped his shoulder. 'She's gone,' she said, her voice filled with a terrible sadness. 'And so will we be if we don't get away from here!' she added more urgently. 'Fang tells me Firestorm sent us a message as he and Nolita fought to reach the well. Nolita wanted us to know that she had no fear of death. Her fears were of physical things and Firestorm was proud to help her with her brave plan.'

Aurora, Shimmer and the day dragons were forcing the night dragons back by weight of numbers, but their fight had turned very ugly. Most of the dragons had already sustained wounds. Fang suddenly attacked out of the darkness, taking one of the flanking night dragons by surprise. But, finding themselves unexpectedly outnumbered, the night dragons were putting up a ferocious fight and the conflict was edging ever closer to the Oracle's well.

Elian allowed Kira to drag him around the well to the far side, but he could not stop staring into the pitch darkness below. He half expected to see Firestorm and Nolita fly back up out of the hole at any moment, but he knew in his heart it would not happen.

The first distant rumble sounded almost like a part of the mêlée that raged at the base of the main ramp. There was no mistaking the second, though. The floor of the chamber shook with the force of it. At the back of the cavern an enormous stalactite broke free from the roof and shattered on impact. All around the cavern, the battling dragons paused and looked as one towards the Oracle's well.

With a suddenness that took Elian totally by surprise, the light level in the chamber brightened as it had when the Oracle had arrived in the past. This time, however, it brightened until it felt as if they could be sitting in the afternoon sun. There was no breathy sighing sound, or misty rising smoke. One heartbeat there was nothing. The next a roaring blue column of fire erupted from the well.

'HOLD!'

The voice was like thunder. It rolled around the chamber with a force that neither man, nor dragon would dare disobey. The air was alive with the smell of power.

'THERE WILL BE NO FURTHER FIGHT-ING. NIGHT DRAGONS, RETURN TO YOUR ENCLAVE.'

The blue column of fire twisted and expanded, racing out through the exit tunnel and, as Elian and Kira were later to discover, up into the sky high above the mountains of Orupee where it gave a similar order to the dragons still fighting in skies. In that moment all conflict ceased. No dragon could deny the Oracle in the fullness of its power.

The column of flame retracted again until it hung above the well. A face resolved in the fire – a face that was both Firestorm and Nolita at the same time.

'Nolita? Are you . . .' Elian began.

'*Dead?*' the Oracle completed in his mind. '*No, Elian, Nolita is very much alive. Her human body died. That was a necessary part of her transformation, but she is very much alive in the fusion that makes me who I am. She thought she would never understand Firestorm, but now we are one. The decision to sacrifice themselves was made by both rider and dragon together – finally Nolita found peace with her destiny.*'

Tears formed in Elian's eyes and began to trickle down his cheeks.

'But you are . . . I am . . . I'm supposed to be you,' he stammered.

'*Do not feel you have failed, Elian,*' the Oracle said kindly. '*You have not failed. Feel rather that Nolita and Firestorm exceeded their life purpose. I did not foresee this fusion, but it has worked out well. The Great Quest was successful. I have been reborn. And I vow to you that I shall do my best to see dragonkind flourish in Areth. Aurora will inevitably feel unfulfilled, as her purpose was overtaken by the actions of your companions. Be ready. I will call on you again in time with a challenge that will satisfy both of your needs.*'

'It is over, then?'

'*Yes, Elian. It is over, for now,*' the Oracle confirmed. '*Go in peace and enjoy some time with your dragon.*'

Chapter Twenty-Seven
Aftermath

'You killed Segun, didn't you?'

'No,' Pell replied, his voice emotionless as he regarded the speaker. 'I would have done it, but another man beat me to the killing blow and I vow I shall never forgive him for it. Shadow finished Widewing, though. She's resting in the trees over there, but beware. She is ready to fight again if you make it necessary.'

Pell's muscles were tense as he watched the man pause on the other side of the grave he had made. He remembered this man. He was one of Segun's lieutenants. The senior night dragonrider looked at the small cairn that Pell had built over Segun's body at the base of the scree slope. He seemed thoughtful. Pell stood poised, his right hand resting on the hilt of the knife he had taken from the night dragonleader.

Despite having his entire enclave set against him, he and his three fellow questors had succeeded. The Great Quest was complete and the Oracle restored to power, but the personal cost for this achievement had been huge. He was surprised the man had not attacked him on sight. As an outlaw to the night dragon enclave, that was what Pell expected when the black dragon swept down towards him.

'You are honest,' the rider said thoughtfully. 'Young and a bit naïve, but strong and honest. With experience and maturity you'll make a fine leader of men. Your ambition was obvious when you approached Segun with news of the Great Quest. We all saw it. Segun hated you from the moment he set eyes on you. I think he saw in you a future rival. The enclave has lost a lot of good men and dragons today. The day dragons took us by surprise with their unorthodox fighting tactics, leaving the night dragon enclave significantly weakened. Worse, the Oracle is rejuvenated and knows full well that virtually our entire enclave tried to prevent that from happening.'

'What's your point?' Pell asked 'I don't see why you're telling me all this. By rights you should have killed me the moment you set eyes on me.'

'True,' the man admitted, his eyebrows drawing together in a deep frown. 'But, given the change in circumstances the enclave now faces, I think your outlaw status should be set aside. You are the only night dragonrider in Areth whom the Oracle is likely to trust. We need you at the enclave, Pell. The Oracle will, no doubt, punish the night dragons for many years to come unless we show a change of heart. Having you with us will speed up that process.'

'You're inviting me back?' Pell exclaimed. 'In what capacity? Surely you're not expecting me to replace Segun? That would be a mockery and you know it. If you're looking for a puppet to manipulate, I'm not interested.'

'No,' the man said firmly. 'You would not be credible as leader of the night dragon enclave. Not yet. One day maybe, but that will depend on how you and your dragon develop. You will need to build a reputation for leadership and strength if you want to lead us. The night dragon enclave has always respected those qualities. It has fallen to me to take on the role for now. I am Korath, rider of Midnight Warrior. And, in my capacity as the new leader of the enclave, I hereby reinstate you. Will you come with me?'

Pell's stomach tightened and the burn of adrenalin swept through his core. Was this a trick? No. He could see Korath was serious. He could return to the enclave. He would be needed and respected. It was more than he could have hoped for.

'Yes,' he said, working hard to keep his voice steady. 'Yes, I'll come. But do we have to leave straight away? The otherworlder, Jack Miller, stole my chance to kill Segun. I want to pay him back for his interference in my affairs.'

'Your revenge will have to wait,' Korath replied. 'The enclave needs you now. Come. The remaining senior riders are waiting for us.'

Pell nodded. 'Very well,' he agreed solemnly. *'We're going back to the enclave,'* he informed Shadow silently. *'Are you ready to fly again?'*

'I am.'

'Good. We want to put on a show of strength for the new leader of the night dragon enclave.'

'Do not worry, Pell. I will not let you down.'

'I know you won't,' he replied, his heart warming at his dragon's bravery. 'As for Jack Miller ...' he added under his breath, '... if I ever chance across him again, he'll wish he'd never met me at all.'

'Ah! Jack Miller. Welcome.'

The Oracle's voice was unlike anything Jack had ever experienced. He could hear it with his ears, yet somehow it reverberated inside his mind as well. The shape of the great dragon's head in the burning

332

column of flame reminded him of something ... someone he had met. Although he had not had many meetings with dragons until these past few days, there was something hauntingly familiar about the creature.

Jack walked alongside Barnabas until they were about ten paces from the Oracle's pit. Barnabas stopped and bowed. Jack followed suit.

'It pleases me that you have come, Jack, for I would like to thank you for your part in my rebirth.'

'I'm afraid I don't know how to address you, Oracle of the dragons,' Jack replied. 'The young people you sent to my world saved my life. What I did was trivial by comparison, but I'm pleased to see them have the success they deserve.'

'Nevertheless, you have my thanks,' the Oracle said. *'But your arrival is also most timely, for there is another who has been waiting to make your acquaintance.'*

'Another?' Jack asked. 'Who could possibly know me here?'

'Jack?' a female voice asked, sounding unsure, yet excited. *'I've searched all over Areth for you, Jack, yet you have eluded me until now. Where have you been hiding?'*

Jack did not know what to say. He looked around the Oracle's cave, but he could not see the owner of the voice. She sounded lovely ... and hauntingly familiar. Then it dawned on him. He had heard the Oracle's voice both in his ears and in his mind, but this voice had rung only in his mind. He spun round, his eyes coming to rest on a day dragon that was approaching down the ramp from the entrance of the Oracle's cave. She was the owner of the voice. He did not know how he knew it, but he could not be more certain.

'Hello?' he said tentatively. 'This is dashed strange! Are you sure you've got the right man? Did you know I'm not from this world?'

'I did not know, Jack,' the dragon replied. *'But that does explain why I could not find you before now. It does not matter where you come from. You are here now. I am Bright Flame, your dragon.'*

Elian looked across the campfire at Kira. The design she had painted on her face this evening reflected brightly in the orange light of the flames. It had been her suggestion to come back to the site in the woods near the meadow where Kasau and his hunters had attacked them. Elian was pleased to be visiting it one last time, but it felt strange to be there without Nolita and Pell.

There had been no sign of Pell after the battle. Jack had told them what he had seen and how he had intervened in Pell's fight with Segun, and the new Oracle had assured them that Pell and Shadow were alive and well, but would not say where they had gone.

'So what are you going to do now, Kira?' Elian asked.

Kira stared intently at the flames and did not look up at the question. 'I'm not entirely sure,' she admitted slowly. 'When I first set out with Fang, I thought I'd be able to rejoin my tribe and take up my place with the hunters when the quest was done, but now I know that can never happen.' She poked at the base of the fire with a stick, causing several clouds of bright sparks to rise. 'Fang wants to take me to the dusk dragon enclave. It's a long way to Ratalucia without the luxury of a short cut through France. The journey will give us a chance to get to know one another better. Now that I've accepted I'll be spending the rest of my life with him, it feels important that I understand more of Fang and how he thinks. How about you?'

'At first I thought I'd go back and see my parents,' Elian admitted sheepishly. 'To let them know I'm all right and tell them about the Great Quest.'

'But you've changed your mind?'

'Yes,' he said, smiling fondly as he glanced at Aurora, who was sleeping nearby. 'There'll be plenty of time for that later. Ra and I are going back to the dawn dragon enclave to spend some time with Tarl, Neema, Blaze and Shimmer. But we're going to fly the long way there this time. I've had my fill of travelling between worlds. It's a bit strange really, because before Aurora came along I was always dreaming about going on adventures and especially about flying on a dragon. I'm sort of hoping this journey will be more like the adventures I dreamed about – exciting but a lot less dangerous!'

Kira laughed. 'I can understand that,' she said. 'You've grown during the past few weeks, Elian. I'm not sure your parents will recognise the boy who they waved goodbye to when you finally decide to return.'

'Oh, I don't know about that . . .'

'I'm serious,' she insisted. 'You've changed more than you realise.'

'I'm going to miss you, Kira,' he countered. 'If I've changed, it's because I've learned a lot of things from you. Thanks.'

Kira could not ignore his gaze any longer. She looked up and was not surprised to see tears welling in Elian's eyes.

'Don't worry, Elian. We'll cross paths again,' she assured him. 'Just you wait and see.'

Sweat ran down Torgan's face, dripping in a steady flow from his nose and chin, as he stood over the anvil hammering the bar of red-hot iron again and again. His huge forearms made light work of hefting the heavy hammer, but his sweat was not from effort so much as from the heat of the forge. There was a bitter wind blowing outside. It had swirled in through the open door of his smithy, time and again through the day, but the blasting heat that was pouring from the forge had pulled its teeth.

The blacksmith continued hammering until the glow in the metal faded. It was almost the right shape now. He returned it to the intense heat of the fire one last time, resting it amongst the coals and then withdrawing the tongs.

Placing the hammer down on the anvil, he laid the tongs next to it and waited for the metal to heat up again. There was a sweat rag on the nearby bench. He turned towards it, but as he reached out his arm a movement near the doorway caught his eye. A strange blur hung in the air. It looked like a miniature cloud of sooty-grey smoke, but it was not escaping the smithy – it was entering.

Torgan picked up the rag and wiped it across his forehead, being careful not to obscure his vision. He had never seen anything like the cloud before. It was fascinating. Smoke and steam did not form coherent puffs of cloud like this. They dispersed quickly.

Whatever it was, the smudge of dark vapour was not following the natural order of gases.

Despite the strong draft the cloud hovered in the doorway. For the briefest instant, a chill swept through Torgan and he sensed danger – too late. With breathtaking suddenness, the thing swooped and struck. There was no chance for the blacksmith to dodge. It hit him full in the chest with enough force to lift him from his feet and fling him across the smithy. He crashed to the floor, stunned.

For a moment he remained there as his hands automatically felt around the impact point on his chest for injury. Strangely, he felt no pain. As he climbed back to his feet, he felt invigorated and stronger than ever before. What had happened? His head felt light. New thoughts were running through his mind. Unusual thoughts. Alien. There were things he had to do. Places he had to visit. He climbed to his feet and walked to the open door. Someone waved at him from across the lane. He waved back. It was a reflex action, but it felt wrong. He did not belong here.

Walking back into the smithy, he went to the very back and opened the cupboard where he kept the few weapons he had made and kept over the years. There was the long sword and the double-bladed axe.

'*Not ideal*,' he thought, taking a careful look at each weapon. '*But it's a start.*'

Crossing the smithy to the door that led into the main part of the house, he opened it and went through into the kitchen. His wife was chopping vegetables for lunch.

'Is everything all right, dear?' she asked. 'You're not finished already, are you?'

'Something's come up,' he replied, his voice sounding strange even in his own ears. 'I've got to go out. I might be gone a while.'

'What are you carrying that for?' she said, pointing at the axe with a horrified expression on her face. 'Are you in some sort of trouble? You're not intending to fight, are you?'

'I'm not in any sort of trouble,' he assured her, moving towards the door that led to their sleeping room. 'There's just something I have to do, that's all.'

She walked to intercept him, raising a hand to stop him from leaving the room before finishing their conversation.

'You're not telling me everything, Torgan,' she accused. 'I can hear it in your voice. And what have you done to your eye? Have you burned it, or something?'

'Which eye?'

'Your left,' she said, moving closer still. 'The colour is different – it looks darker. Are you sure you're all right?'

'I'm fine, dear,' he said. 'Everything is just fine.'

Author's Note
on Baron Manfred von Richthofen

An aura of mystery shrouds the death of Baron Manfred von Richthofen. In writing about it here, I tried to recreate the path and timing of his last flight as accurately as possible from the information available. It has been fairly well established that his death was caused by machine gun fire from the ground. The bullet that killed him was a standard .303 round fired from long range, but as to who fired the fatal shot – I'm not sure that we will ever know for sure. On balance of probability it seems likely that Sgt C. B. Popkin, a Vickers gunner with the 24th Machine Gun Company, had the best claim. However, it is said that even he had his doubts. Who knows? Maybe there was a dusk dragon there after all!

Regardless of who did fire the killing shot, one

thing is clear – 21st April 1918 shall long be remembered as the day the world of aviation lost one of its most colourful and charismatic characters.

FOUR DRAGONRIDERS ON A MISSION TO SAVE THEIR WORLD

FIRESTORM
Book 1 in the fantasy series.

Nolita is terrified of dragons! Learning to fly her day dragon is dangerous enough without irrational fears to contend with and a vicious dragonhunter on her tail. With Elian, another novice rider, she seeks the first of four orbs, to save the leader of all dragonkind. To do so, she must face her worst fears, and face them alone ...

ISBN: 978-1-84738-068-5

SHADOW
Book 2 in the fantasy series.

Pell and his night dragon Shadow must find the dark orb to help save the Oracle, leader of all dragonkind. But Segun, a power-hungry tyrant, stands in their way. Pell must use his flying skills, bravery and resourcefulness to the limit, as Segun is determined to get the orb - even if it means killing the opposition.

ISBN 978-1-84738-069-2

DRAGON ORB: LONGFANG

Book 3 in the fantasy series.

FOUR DRAGONRIDERS ON A MISSION TO SAVE THEIR WORLD

Kira and her dusk dragon, Longfang, must find the third orb to save the Oracle, leader of all dragonkind. Following a path beset with dangers, and traps that maim and kill, the four dragonriders must reach the twilight world of the Castle of Shadows. Kira knows enough to be anxious. What twisted sacrifice will this orb demand?

ISBN: 978-1-84738-070-8